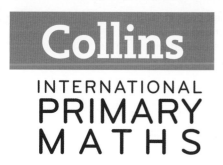

Collins

INTERNATIONAL PRIMARY MATHS

Student's Book 4

William Collins' dream of knowledge for all began with the publication of his first book in 1819. A self-educated mill worker, he not only enriched millions of lives, but also founded a flourishing publishing house. Today, staying true to this spirit, Collins books are packed with inspiration, innovation and practical expertise. They place you at the centre of a world of possibility and give you exactly what you need to explore it.

Collins. Freedom to teach.

Published by Collins
An imprint of HarperCollinsPublishers
The News Building
1 London Bridge Street
London
SE1 9GF

HarperCollinsPublishers
Macken House,
39/40 Mayor Street Upper,
Dublin 1,
D01 C9W8, Ireland

Browse the complete Collins catalogue at
www.collins.co.uk

© HarperCollinsPublishers Limited 2021

10 9 8 7 6 5

ISBN 978-0-00-836942-2

British Library Cataloguing-in-Publication Data
A catalogue record for this publication is available from the British Library.

Author: Caroline Clissold
Series editor: Peter Clarke
Publisher: Elaine Higgleton
Product developer: Holly Woolnough
Project manager: Mike Harman (Life Lines Editorial Services)
Development editor: Joan Miller
Copyeditor: Tanya Solomons
Proofreader: Tanya Solomons
Cover designer: Gordon MacGilp
Cover illustrator: Ann Paganuzzi
Typesetter: QBS Learning
Illustrators: Ann Paganuzzi and QBS Learning
Production controller: Lyndsey Rogers
Printed and bound in India by Replika Press Pvt. Ltd.

With thanks to the following teachers and schools for reviewing materials in development: Antara Banerjee, Calcutta International School; Hawar International School; Melissa Brobst, International School of Budapest; Rafaella Alexandrou, Pascal Primary Lefkosia; Maria Biglikoudi, Georgia Keravnou, Sotiria Leonidou and Niki Tzorzis, Pascal Primary School Lemessos; Taman Rama Intercultural School, Bali.

MIX
Paper | Supporting
responsible forestry
FSC
www.fsc.org   FSC™ C007454

This book is produced from independently certified FSC™ paper to ensure responsible forest management.

For more information visit: www.harpercollins.co.uk/green

The publishers gratefully acknowledge the permission granted to reproduce the copyright material in this book. Every effort has been made to trace copyright holders and to obtain their permission for the use of copyright material. The publishers will gladly receive any information enabling them to rectify any error or omission at the first opportunity.
Cambridge International copyright material in this publication is reproduced under licence and remains the intellectual property of Cambridge Assessment International Education

Contents

Number

Geometry and Measure

Statistics and Probability

How to use this book

This book is used towards the start of a lesson when your teacher is explaining the mathematical ideas to the class.

- An **objective** explains what you should know, or be able to do, by the end of the lesson.

Key words

- The **key words** to use during the lesson are given. It's important that you understand the meaning of each of these words.

Let's learn

This section of the Student's Book page **teaches** you the main mathematical ideas of the lesson. It might include pictures or diagrams to help you **learn**.

1. An activity that involves thinking and working mathematically.

An activity or question to discuss and complete in pairs.

Guided practice

Guided practice helps you to answer the questions in the Workbook. Your teacher will talk you through this question so that you can work independently with confidence on the Workbook pages.

HINT

Use the page in the Student's Book to help you answer the questions on the Workbook pages.

1. **Thinking and Working Mathematically** (TWM) involves thinking about the mathematics you are doing to gain a deeper understanding of the idea, and to make connections with other ideas. The TWM star at the back of this book describes the 8 ways of working that make up TWM. It also gives you some sentence stems to help you to talk with others, challenge ideas and explain your reasoning.

At the back of the book

Lesson 1: **Counting in 10s, 100s and 1000s**

* Count on and back in steps of 10, 100 and 1000

Key words
* zero
* ones
* tens
* hundreds
* thousands

Let's learn

Look at the signpost.

The distance to Quebec City is 727 kilometres.

What if it was 1000 km further away?

The distance to Moscow is 7492 kilometres.

What if it was 300 km closer?

What is the distance to Tokyo?

What is the distance to Riyadh?

What is the distance to Taipei?

Draw place value counters to represent the distance to Ulaanbaatar.

Now represent the number that is 1000 **more**. Write it down and say it to your partner.

Represent the number that is 300 more. Write it and say it.

Represent the number that is 80 more. Write it and say it.

Now represent the number that is 4 less than your last number.

What is your new number? Write it in words and numerals.

Guided practice

Write the missing numbers as you count on or back in 1000s.

1246, 2246, | 3246 |, | 4246 |, | 5246 |, | 6246 |, 7246

8504, 7504, | 6504 |, | 5504 |, | 4504 |, | 3504 |, | 2504 |

Lesson 2: Adding odd and even numbers

Key words
- even number
- odd number
- add
- sum
- total

Number

- Recognise and explain rules for adding odd and even numbers

Let's learn

An even number is a multiple of 2. An odd number is 'an even number add 1'.

$4 = 2 + 2$ $6 = 2 + 2 + 2$ $3 = 2 + 1$ $5 = 4 + 1$ $7 = 6 + 1$

There are rules for adding even and odd numbers.

 even number + even number = even number

 odd number + odd number = even number

 even number + odd number = odd number

Guided practice

Even numbers		Not even numbers	
468	574	567	479
350	902	843	671

Choose pairs of numbers from the Carroll diagram.

Write whether they will give an even or odd sum.

$468 + 567 =$ odd sum

$350 + 902 =$ even sum

$567 + 843 =$ even sum

$574 + 479 =$ odd sum

 Can you say if the sum of two numbers is odd or even without calculating?

Is the sum of these numbers odd or even? $3456 + 2758$

What about the sum of these numbers? $8165 + 6239$

Talk to your partner about how you know.

Number

Lesson 3: **Subtracting odd and even numbers**

Key words
- **even number**
- **odd number**
- **subtract**
- **difference**

- Recognise and explain rules for subtracting odd and even numbers

Let's learn

Like addition, there are rules for subtracting even and odd numbers.

even number – even number = even number

E – E = E
6 – 4 = 2

odd number – odd number = even number

O – O = E
5 – 3 = 2

even number – odd number = odd number

E – O = O
8 – 1 = 7

odd number – even number = odd number

O – E = O
9 – 6 = 3

Is the difference between 3456 and 2349 odd or even? How do you know?

Write some calculations that have an odd difference.

How do you know that the differences will be odd?

Write some calculations that have an even difference.

How do you know that the differences will be even?

Guided practice

Choose pairs of numbers from the Carroll diagram.

Will they give an even or odd difference?

Even numbers		Not even numbers	
3262	3264	3261	3263
3266	3268	3265	3267

3262 – 3261 = odd difference

3268 – 3264 = even difference

3265 – 3262 = odd difference

3267 – 3263 = even difference

Lesson 4: **Sequences**

• Recognise and extend number sequences

Key words
• sequence
• pattern
• increase
• decrease
• term
• rule

Number

Let's learn

The house numbers on this side of the street increase by 2 each time.

They make a number sequence.

Number sequences are patterns of numbers that follow rules.

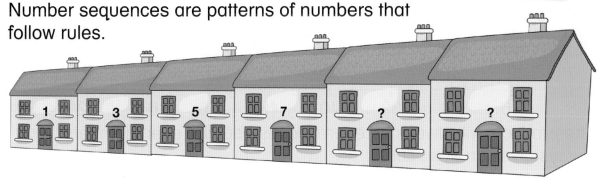

Sequences can show numbers increasing (getting larger) or decreasing (getting smaller).

Each number in a number sequence is called a **term**.

You need to know three or four of the terms in the sequence to work out the rule.

The terms in this sequence increase by 2 each time (+2).

2, 4, 6, 8, 10…

It is also a sequence of the multiples of 2.

Make up a number sequence with the rule +3.

What is the first number of your sequence? What is the 10th number?

Guided practice

This number sequence is based on the multiples of a number.
Which times table does the sequence show?

56, 49, 42, 35, 28   These are all multiples of ⟨ 7 ⟩.

The rule is | subtract 7 from the previous term |.

Lesson 1: **Counting in 1-digit steps**

Number

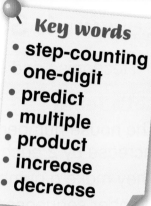

Key words
- step-counting
- one-digit
- predict
- multiple
- product
- increase
- decrease

- Count on and back in 1-digit steps of constant size

Let's learn

Counting from 0 in 8s gives the multiples of 8 in the 8 times table.

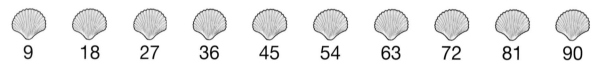

1	2	3	4	5	6	7	8	9	10
8	16	24	32	40	48	56	64	72	80

The products will always be even because adding even numbers always gives an even sum.

What do you notice about counting in 9s?

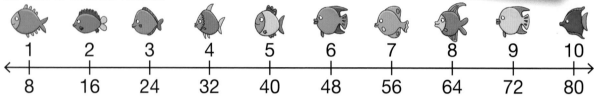

| 9 | 18 | 27 | 36 | 45 | 54 | 63 | 72 | 81 | 90 |

Look for patterns in the multiples of 8 on the number line above.

Write the numbers you say when you count in 4s.

Write the numbers you say when you count in 2s.

Now draw a ring around the numbers that you say when you count in 2s, 4s and 8s.

What do you notice?

Guided practice

Count on in steps of 5 from 0 to 50.

0, $\boxed{5}$, $\boxed{10}$, $\boxed{15}$, $\boxed{20}$, $\boxed{25}$, $\boxed{30}$, $\boxed{35}$, $\boxed{40}$, $\boxed{45}$, 50

Use the pattern you can see to count on in 5s from 1 to 51.

1, $\boxed{6}$, $\boxed{11}$, $\boxed{16}$, $\boxed{21}$, $\boxed{26}$, $\boxed{31}$, $\boxed{36}$, $\boxed{41}$, $\boxed{46}$, 51

What patterns do you notice?

The numbers always have a difference of 5.
The numbers are odd and then even in the pattern.

Lesson 2: **Counting in 1-digit steps beyond zero**

Key words
- positive
- negative
- increase
- decrease

- Count on and count back in 1-digit steps beyond zero

Let's learn

Numbers can be positive or negative.

Positive numbers are all greater than zero.

Negative numbers are all less than zero. We always write a negative sign (–) in front of a negative number.

Negative numbers mirror positive numbers across zero!

$$-10 \; -9 \; -8 \; -7 \; -6 \; -5 \; -4 \; -3 \; -2 \; -1 \; 0 \; 1 \; 2 \; 3 \; 4 \; 5 \; 6 \; 7 \; 8 \; 9 \; 10$$

Put your finger on 10 and count back 15. Did you land on –5?

Put your finger on –10 and count on 18. Did you land on 8?

Count on in steps of 10 from 0. What is the 7th step?

Count back in steps of 10 from 0. What must the 7th step be?

Count on in steps of 8 from 0. What is the 9th step?

Count back in steps of 8 from 0. What must the 9th step be?

Count on in steps of 3 from 0. What is the 8th step?

Count back in steps of 3 from 0. What must the 8th step be?

Guided practice

Matsu counts on in steps of 8 from 0. These are the numbers she says: 0, 8, 16, 24, 32, 40, 48, 56, 64, 72, 80

She then counts back in steps of 8 from 0.

What numbers does she say?

–8, –16, –24, –32, –40, –48, –56, –64, –72, –80

Lesson 3: **Sequences**

Number

• Recognise and extend number sequences

Let's learn

Number sequences are made by adding, subtracting, multiplying or dividing the previous number.

12,   15,   18,   21,   24,   27

The rule for this number sequence is 'Start on 12. Add 3 to the previous term.'
The next three terms are 30, 33, 36.
What are the next three terms in this sequence?

The numbers in a sequence are called **terms**.

45,   36,   27,   18,   9,   0,   −9,   −18

The rule is 'Start on 45, subtract 9.'

The next three terms are −27, −36, −45.

Look at this sequence.

1000,   200,   40,   8

The rule is 'Start on 1000, divide by 5.'

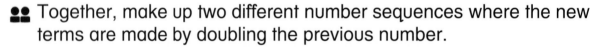 Together, make up two different number sequences where the new terms are made by doubling the previous number.

Now make up a number sequence where the new terms are made by halving the previous number.

Guided practice
What are the rules for these sequences?

a 2, 4, 8, 16, 32

Double the previous term.

b 256, 64, 16, 4

Divide the previous term by 4.

Lesson 4: **Square numbers**

- Recognise and extend the pattern of square numbers

Number

Let's learn

A number that can be represented by a square array is a **square number**.
Here are the first six square numbers.

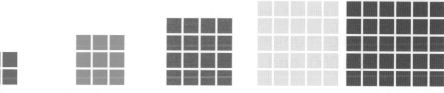

1 × 1 = 1   2 × 2 = 4   3 × 3 = 9   4 × 4 = 16   5 × 5 = 25   6 × 6 = 36

Here is another way to represent square numbers:

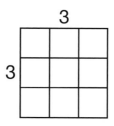
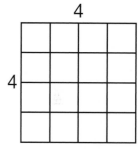

1 × 1 = 1   2 × 2 = 4   3 × 3 = 9   4 × 4 = 16

With your partner, make a poster to show square numbers up to 36.

Draw grids like the ones above for all square numbers up to 6 × 6.

Cut them out. Stick them on poster paper.

Label each side with the number of squares and write the times table fact.

Guided practice

Draw diagrams to show these square numbers.

Label each diagram with the times table fact that shows the square number.

1  1 × 1 = 1   25 5 × 5 = 25

Number

Lesson 1: **Reading and writing numbers to 1000**

- Read and write numbers to 1000

Key words
- **numeral**
- **number**
- **hundreds**
- **tens**
- **ones**
- **digit**
- **place holder**
- **zero**

Let's learn

814   eight hundred and fourteen

There are 8 hundreds (800), 1 ten (10) and 4 ones (4).

It is a 3-digit number so it is less than 1000, and it is even.

909   nine hundred and nine

There are 9 hundreds (900), no tens (0) and 9 ones (9).

It is a 3-digit number so it is less than 1000, and it is odd. The zero in the tens position means there are no tens. Zero is a place holder.

Use these words to make up some 3-digit numbers.

1 **six   eight   seven   hundred   and   fifty**

How many different numbers can you make?

Which number has been the same each time? Why?

Which numbers have changed?

Explain to your partner how and why they have changed.

Guided practice

Put the words together to make two different 3-digit numbers.

twenty   four   hundred   and   six

> four hundred and twenty-six, six hundred and twenty-four

nine   five   seventy   hundred   and

> nine hundred and seventy-five, five hundred and seventy-nine

Number

Lesson 2: **Reading and writing numbers to 10 000**

• Read and write numbers to 10 000

Let's learn

5246

five thousand, two hundred and forty-six

| 1000 | 1000 | 1000 | 1000 | 1000 | 100 | 100 | 10 | 10 | 10 | 10 |

| 1 | 1 | 1 | 1 | 1 | 1 |

In this number there are 5 thousands, 2 hundreds, 4 tens and 6 ones.

| 1000 | 1000 | 1000 | 1000 | 1000 | 1000 | 1000 | 1000 | 1000 | 100 | 100 | 100 | 100 | 100 | 1 | 1 | 1 |

In the number 9503 there are 9 thousands, 5 hundreds and 3 ones.
There are no tens. I can show this by writing a zero.

9503

I can write the number 9503 as nine thousand, five hundred and three.

Choose a 4-digit number.
Draw place value counters to represent your number.
Talk to each other about what your number is and how it is made up.
Write your number as a numeral and in words.

Guided practice
Complete the number.

3705: three ⎢ thousand ⎢, seven ⎢ hundred ⎢ and ⎢ five ⎢

15

Lesson 3: **Reading and writing numbers to 100 000**

- Read and write numbers to 100 000

Let's learn

This number is 25 246.

It is written as twenty-five thousand, two hundred and forty-six.

There are 25 thousands, 2 hundreds, 4 tens and 6 ones.

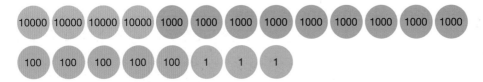

This number is 49 503.

It is written as forty-nine thousand, five hundred and three.

This time there are 49 thousands, 5 hundreds and 3 ones. There are no tens. In numerals, this is shown by a zero.

Choose a 5-digit number.
Draw place value counters to represent your number.
Talk to each other about what your number is and how it is made up.
Write your number as a numeral and in words.

Guided practice
Write the number in numerals.

eighty-three thousand, four hundred and thirty-seven   83 437

Lesson 4: **Negative numbers**

• Read, write and count with negative numbers

Key words
• **numeral**
• **number**
• **tens**
• **ones**
• **digit**
• **zero**
• **positive number**
• **negative number**

Number

Let's learn

Look at this number line. There are numbers on both sides of zero.

The red numbers are positive numbers.
The blue numbers are negative numbers.

We read and say 'negative five'.

–10 –9 –8 –7 –6 (–5) –4 –3 –2 –1 0 1 2 3 4 5 6 7 8 9 10

⟵————————Negative numbers Positive numbers————————⟶

Negative numbers describe very low temperatures, depths below sea level and underground floors in buildings. They can also be used for money.

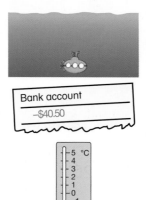

The submarine is 300 m below sea level. We write this as –300 m.

Sean's owes the bank $40.50.

The temperature is negative two: it's very cold.

Take turns to choose a number on the number line. Give your partner instructions, such as:

Put your finger on 3, count back 10, count forwards 5. Where are you?

Guided practice

Draw a line from each number to show where it belongs on the number line.

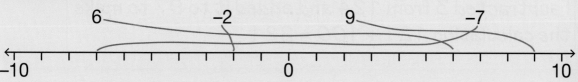

6   –2   9   –7

⟵——————————————————————————⟶
–10   0   10

Lesson 1: **Mental addition**

Number

• Use mental calculation strategies to add

Key words
• **addition**
• **augend**
• **addend**
• **total**
• **mental calculation strategies**
• **rounding and adjusting**

Let's learn

When you add, always try to use a mental calculation strategy first.

$68 + 59 =$

First, **estimate** the sum. 68 is close to 70. 59 is close to 60. A good estimate is 130.

Now **calculate** the sum. Take 1 from 68 and add it to 59. This gives $67 + 60$. Then double 60 and add 7.
$68 + 59 = 67 + 60 = 127$

Now we must **check**. We can add the numbers the other way around because addition is commutative.

$\bigcirc + 10 = 30$.

We can use our knowledge of the relationship between addition and subtraction to find the value of the circle. We subtract 10 from 30 to give 20. The value of the circle is 20.

Talk to your partner about how you could add these numbers.

$245 + 134 =$

$356 + 213 =$

Guided practice

Use a mental strategy to add these numbers. Explain your strategy.

$124 + 97 = \boxed{221}$

I subtracted 3 from 124 and added it to 97 to make the calculation $121 + 100 = 221$

Lesson 2: **Mental addition involving money**

- Use mental calculation strategies to add money

Key words
- dollar
- cent
- augend
- addend
- total
- partitioning
- mental calculation strategy
- rounding and adjusting

Let's learn

Partitioning is another mental calculation strategy that we can use.

$368 + $411 =

> First, **estimate** the total cost. $368 is close to $400. $411 is close to $400. A good estimate is $800.

> Next, **calculate** the total cost. Keep $368 whole and partition $411, then add each part.
> $368 + $400 + $10 + $1 = $779

> Now we must **check**. I am going to add the numbers the other way round.

☐ + △ = £45.

The two shapes can represent any two different numbers. They must add up to £45.

👤 A computer game costs $145 . How much would two cost?

7 What mental calculation strategy can we use to find the total cost?

Talk to your partner about how to do this.

Work out the total cost.

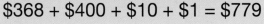

Guided practice

Use a mental strategy to add these amounts. Explain your strategy.

$675 + $98 = $773

I took $2 from $675 and added it to $98 to make
$673 + $100 = $773

Lesson 3: **Mental subtraction**

Number

- Use mental calculation strategies to subtract

Key words
- subtraction
- minuend
- subtrahend
- difference
- counting on
- counting back

Let's learn

Counting on is a useful mental strategy, especially if the numbers are close together.

$543 - 487 =$

Estimate first. 50 is a sensible estimate, as $550 - 500 = 50$.

Then, count on or back along a number line to find the difference.

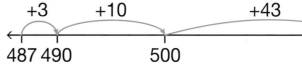

$3 + 10 + 43 = 56$

The difference between 543 and 487 is 56.

We could check by counting back.

👥 Write two 3-digit numbers.

Draw a number line and put the smaller number on the left and the greater number on the right.

Find the difference by counting on from the smaller number.

Your partner checks the difference by counting back.

Guided practice

Use a mental strategy to subtract these numbers. Explain your strategies.

a $375 - 99 = \boxed{276}$

I rounded 99 to 100 and subtracted and then added 1.
$375 - 100 + 1 = 276$

b $125 - 98 = \boxed{27}$

I counted on from 98 to 100 and then to 125. The difference is 27.

Number

Lesson 4: **Mental subtraction involving money**

- Use mental calculation strategies to subtract money

Key words
- **dollar**
- **cent**
- **minuend**
- **subtrahend**
- **difference**
- **partitioning**
- **rounding**

Let's learn

We can use rounding and adjusting to subtract.

Masa has seen the same set of toy animals in two shops. She wants to know the price difference between them.

Round $28 to $30. $39 – $30 = $9. We have subtracted $2 too many so we must add $2.

The difference in price is $11.

Now we need to check.
We can do this by adding $11 onto $28 to see if that makes $39.

Gift Grab $28

Toy World $39

$\bigcirc - \$10 = \$20$

To find the missing number we could add $10 and $20. That means the circle is worth $30.

Subtract $99 from each of these amounts.

$256 $398 $478

Check your partner's answers by adding $99 onto the differences.

Guided practice

Use a mental strategy to subtract these amounts. Explain your strategy.

$345 – $98 = $247

I subtracted $100 and added $2. $345 – $100 + $2 = $247

Lesson 1: **Adding 3-digit and 2-digit numbers (1)**

Number

• Estimate and add 3-digit and 2-digit numbers

Key words
• augend
• add
• addend
• equals
• sum
• total
• estimate
• partitioning

Let's learn

$$643 + 55 =$$

(100) (100) (100) (10) (10) (1) (1)   (10) (10) (10) (1) (1) (1)
+
(100) (100) (100) (10) (10) (1)   (10) (10)   (1) (1)

Start with an estimate.

To estimate, we can round to the nearest 10. Multiples of 10 are quite easy to add.
643 is close to 640 and 55 is close to 60. A good estimate would be 700.

We should use a mental strategy if we can.

Partitioning would be a good strategy.

Use a written method to add numbers when the calculation is too hard for a mental strategy.

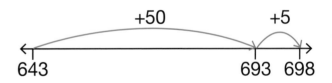

+50   +5

643   693 698

This is the expanded written method.

This is the formal written method.

```
   6 4 3
 +   5 5
       8
     9 0
   6 0 0
   6 9 8
```

```
   6 4 3
 +   5 5
   6 9 8
```

Guided practice
Use partitioning to work out the answer.

$$465 + 24 = \boxed{489}$$

Explain your strategy.

I partitioned 24 and added 20 and then 4 to 465.

👥 Talk to your partner about how partitioning and using place value counters are similar.
How are they different?

Lesson 2: **Adding 3-digit and 2-digit numbers (2)**

Number

- Estimate and add 3-digit and 2-digit numbers

Let's learn

$$646 + 39 =$$

100	100	10	10	1	1		10	10	1	1	1
100	100	10	10	1	1	+	10		1	1	1
100	100			1	1			1	1	1

Start with an estimate. Round the numbers to the nearest 10 and add. 646 is close to 650 and 39 is close to 40. A good estimate would be 690.

Use a mental strategy if you can.

Compensation would be a good strategy. 39 is close to 40, so add 646 and 40. But as 40 is 1 more than 39, we have added an extra 1. We **compensate** by subtracting 1.

```
  6 4 6
+   3 9
  1 5
  7 0
6 0 0
6 8 5
```

+40

−1

646   685 686

This is the expanded written method.

This is the formal written method.

```
  6 4 6
+   3 9
  6 8 5
      1
```

Make 524 and 69 with place value counters.

Estimate the total.

Use compensation to work it out.

Now practise using the two written methods.

Guided practice

Use the compensation strategy to work out the answer. Explain your strategy.

$$354 + 29 = \boxed{383}$$

I added 30 to 354 and then subtracted 1.

Lesson 3: **Adding pairs of 3-digit numbers (1)**

Number

• Estimate and add pairs of 3-digit numbers

Let's learn

$$446 + 325 =$$

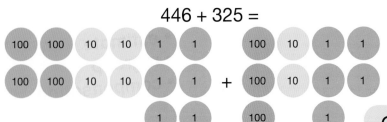

Key words
* augend
* add
* addend
* equals
* sum
* total
* estimate
* regroup

Start with an estimate. We can round 446 to 450 and 325 to 330. A good estimate would be 780.

Once we have simplified the calculation we can use a number line to help us add.

We can use a **making 10** strategy.

$+300$ $\qquad$ $+20$ $\quad$ $+1$

450 $\qquad$ 750 $\qquad$ 770 771

$446 + 325 = 450 + 321 = 771$

```
  4 4 6
+ 3 2 5
  1 1
  6 0
7 0 0
7 7 1
```

Although we could use a mental strategy, it might be easier to use a written method.

This is the expanded written method.

This is the formal written method.

```
  4 4 6
+ 3 2 5
  7 7 1
    1
```

👥 Make 538 and 327 with place value counters.

Estimate the total.

Find it using the making 10 strategy.

Then check by using both written methods.

Guided practice

Use the making 10 strategy to work out the answer.

$346 + 198 =$ $\boxed{344 + 200 = 544}$

Explain your strategy.

> I made a multiple of 10 by subtracting 2 from 346 and adding it to 198. Then I added 344 and 200.

Lesson 4: **Adding pairs of 3-digit numbers (2)**

- Estimate and add pairs of 3-digit numbers

Let's learn

$$367 + 276 =$$

For numbers like this, an expanded or formal written method would probably be best.

Start with an estimate. We could round these numbers to the nearest 100 for our estimate. Round 367 to 400 and 276 to 300. A good estimate would be 700.

We could use a mental strategy to find the sum. I'm not sure which one would be best though.

This is the formal written method.

```
  3 6 7
+ 2 7 6
  6 4 3
  1 1
```

```
  3 6 7
+ 2 7 6
    1 3
  1 3 0
  5 0 0
  6 4 3
```

This is the expanded written method.

We can check by adding the other way around or by subtracting one of the numbers we added from the sum.

Make 365 and 287 with place value counters.

Estimate the total.

Find it using Base 10 equipment.

Then show both written methods.

Once you have the total, check it using a calculator.

Guided practice

Use the formal written method to work out the answer.

```
  3 7 6
+ 2 4 8
  6 2 4
  1 1
```

$$376 + 248 =$$

25

Number

Lesson 1: **Subtracting 2-digit from 3-digit numbers (1)**

• Estimate and subtract 2-digit numbers from 3-digit numbers

Let's learn

486 – 73 =

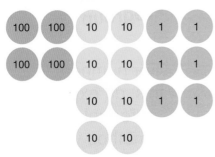

Start with an estimate. We can round 486 to 490 and 73 to 70. We subtract 70 from 490 for an estimate. A good estimate is 420.

We should use a mental strategy if we can. **Partitioning** would be a good method.

Keep 486 whole and subtract 70, then subtract 3.

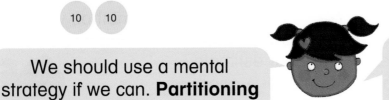

Use a written method to subtract when the calculation is too hard for a mental strategy.

$$\begin{array}{rrr} 400 & 80 & 6 \\ - & 70 & 3 \\ \hline 400 & 10 & 3 \end{array}$$

This is the expanded written method.

This is the formal written method.

$$\begin{array}{r} 4\ 8\ 6 \\ -\ \ \ 7\ 3 \\ \hline 4\ 1\ 3 \end{array}$$

400 + 10 + 3 = 413

👥 Make 798 with place value counters.

Estimate the difference when you subtract 36.

Use a mental calculation strategy to find 798 – 36.

Use Base 10 equipment to subtract 36 from 798.

Now practise the two written methods.

Guided practice
Use a mental strategy to find the difference.

476 – 53 = ⟨423⟩

Explain your strategy.

I used partitioning. I subtracted 50 first and then I subtracted 3.

Number

Lesson 2: **Subtracting 2-digit from 3-digit numbers (2)**

• Estimate and subtract 2-digit numbers from 3-digit numbers

Let's learn

687 – 68 =

Start with an estimate. 68 is close to 70. Subtract 70 from 687 to give an estimate of 617.

We can use the mental strategy called **compensation**.

68 is close to 70, so subtract 70 from 687. But 70 is 2 more than 68, so we have subtracted 2 too many. We **compensate** by adding 2.

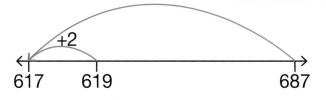
−70
+2
617   619   687

Now check using the written methods. Which do you prefer?

```
        70   17
600   8̶0̶   7̶
 −      60   8
600   10   9
```
600 + 10 + 9 = 619

```
  6⁷8̶¹7
−   6 8
  6 1 9
```

• Make 746 with place value counters.

Subtract 88.

Estimate the difference first.

Use compensation to find the difference.

Check the difference using the two written methods.

Guided practice

Use a mental strategy to subtract.

682 – 79 = 603

Explain your strategy.

I used compensation. I rounded the subtrahend to the next 10. I subtracted 80 and then added 1.

Number

Lesson 3: **Subtracting pairs of 3-digit numbers (1)**

Key words
- minuend
- subtract
- subtrahend
- equals
- difference
- estimate
- regroup

• Estimate and subtract pairs of 3-digit numbers

Let's learn

$$584 - 538 =$$

| 100 | 100 | 10 | 10 | 1 | 1 |

| 100 | 100 | 10 | 10 | 1 | 1 |

| 100 | | 10 | 10 | | |

| | | 10 | 10 | | |

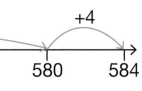

We want to work out 584 – 538.
Start with an estimate.
$$580 - 540 = 40$$

+2   +40   +4

← 538 540   580   584 →

Let's use counting on to work this out. We count on 2 + 40 + 4 from 538 to reach 584. That's 46. So 584 – 538 = 46.

These are the written methods. Which do you prefer?

		70	14		
	500	8̶0̶	4̶		$5^7 8^1 4$
–	500	30	8		– 5 3 8
		40	6		4 6

$40 + 6 = 46$

👥 Make 235 with place value counters.

Subtract 185.

Estimate the difference first.

Show how to find the difference by counting on or counting back along a number line.

Now use the formal written method to work out the answer.

Discuss with your partner which method you think is better.

Guided practice

Use the formal written method to work out the answer.

$$345 - 237 =$$

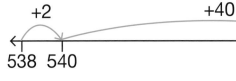

$$3\ ^3\!4\ ^1\!5$$
$$-\ 2\ 3\ 7$$
$$1\ 0\ 8$$

Lesson 4: **Subtracting pairs of 3-digit numbers (2)**

• Estimate and subtract pairs of 3-digit numbers

Let's learn

$624 - 356 =$

We want to subtract 356 from 624.

Start with an estimate.

$600 - 400 = 200$

$$\begin{array}{ccc} \overset{500}{\cancel{600}} & \overset{110}{\cancel{20}} & \overset{14}{\cancel{4}} \\ -\ 300 & 50 & 6 \\ \hline 200 & 60 & 8 \end{array}$$

$200 + 60 + 8 = 268$

This subtraction will need regrouping in the ones and the tens! It might not be easy to use a mental strategy. Let's use a written method.

$$\begin{array}{r} ^5\cancel{6}^1\cancel{2}^14 \\ -\ 3\ 5\ 6 \\ \hline 2\ 6\ 8 \end{array}$$

Make 931 with place value counters.

Subtract 256.

Estimate the difference first.

Find the difference.

Use both written methods to work out the answer.

Check the difference by adding it to the subtrahend.
Do you get the minuend?

Guided practice

Use the formal written method to work out the answer.

$353 - 186 =$

$$\begin{array}{r} ^2\cancel{3}^{14}\cancel{5}^13 \\ -\ 1\ 8\ 6 \\ \hline 1\ 6\ 7 \end{array}$$

Number

Lesson 1: **5 and 10 times tables**

- Understand the relationship between the 5 and 10 times tables

Let's learn

We can use **commutativity** and the **inverse relationship** between multiplication and divison to help us recall the times table facts.

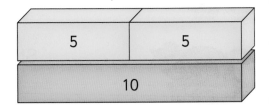

The model shows $5 \times 2 = 10 \times 1$.
10 is double 5 and 5 is half of 10.

This makes multiplying by 5 easy. To multiply 18 by 5, multiply 18 by 10 and then halve the product.

$18 \times 10 = 180$
Half of $180 = 90$
$18 \times 5 = 90$

Write five 2-digit even numbers less than 30.
Multiply each number by 5 by multiplying by 10 and halving.
Talk to your partner. Is this a useful strategy? Why?

Guided practice

Feechi knows that $5 \times 8 = 40$.

What commutative fact does he know? $\boxed{8 \times 5 = 40}$

Write the two inverse facts he also knows.

$\boxed{40 \div 5 = 8}$ $\boxed{40 \div 8 = 5}$

Lesson 2: **2, 4 and 8 times tables**

Key words
- multiply
- product
- divide
- quotient
- double
- half
- commutative
- inverse

Number

- Understand the relationship between the 2, 4 and 8 times tables

Let's learn

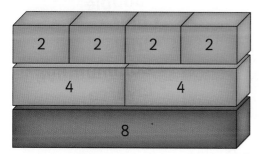

The model shows $2 \times 2 = 4 \times 1$.

4 is double 2 and 2 is half of 4.

We can also see that $4 \times 2 = 8 \times 1$.

8 is double 4 and 4 is half of 8.

So multiplying by 8 is easy.

To multiply 15 by 8, multiply 15 by 2, then double it and double it again.

$15 \times 2 = 30$

Double 30 is 60.

Double 60 is 120.

$15 \times 8 = 120$

Write five 2-digit even numbers that are less than 20.

Multiply each number by 8 by multiplying by 2, doubling and doubling again. You can make jottings.

Talk to your partner. Is this a useful strategy? Why?

Guided practice

Work out the answers. Use the doubling strategy to find the products.

a $18 \times 4 =$

$18 \times 2 = 36, 36 \times 2 = 72$

b $21 \times 8 =$

$21 \times 2 = 42, 42 \times 2 = 84, 84 \times 2 = 168$

Lesson 3: **3, 6 and 9 times tables**

Number

- Understand the relationship between the 3, 6 and 9 times tables

Let's learn

3	3	3	3	3	3	3	3	3	3	3	3

6	6	6	6	6	6

9	9	9	9

We can see from the model that double 3 is 6.

We can double the 3 times table to find facts for the 6 times table. Here is an example:

$12 \times 3 = 36$ $36 \times 2 = 72$

$12 \times 6 = 72$

3	3

6

$3 \times 2 = 6 \times 1$

We can see from the model that triple 3 is 9.

We can triple the 3 times table to find facts for the 9 times table. Here is an example:

$21 \times 3 = 63$ $63 \times 3 = 189$

$21 \times 9 = 189$

3	3	3

9

$3 \times 3 = 9 \times 1$

👥 Use the diagrams of the coloured rods to help you make up four different multiplication statements. For each one, write the commutative fact and the two inverse division facts.

Guided practice

Work out the answer. Use the strategy: multiply by 3 then double.

$25 \times 6 =$

$25 \times 3 = 75, 75 \times 2 = 150$

Lesson 4: **7 times table**

• Know the 7 times table

Key words
• **multiply**
• **product**
• **divide**
• **quotient**
• **commutative**
• **inverse**

Number

Let's learn

This table shows all the multiplication facts we need to know.

The orange column shows the 7 times table.

Because multiplication is commutative, we know most of these facts from the other tables.

×	1	2	3	4	5	6	7	8	9	10
1	1	2	3	4	5	6	7	8	9	10
2	2	4	6	8	10	12	14	16	18	20
3	3	6	9	12	15	18	21	24	27	30
4	4	8	12	16	20	24	28	32	36	40
5	5	10	15	20	25	30	35	40	45	50
6	6	12	18	24	30	36	42	48	54	60
7	7	14	21	28	35	42	49	56	63	70
8	8	16	24	32	40	48	56	64	72	80
9	9	18	27	36	45	54	63	72	81	90
10	10	20	30	40	50	60	70	80	90	100

• $3 \times 7 = 21$, so $7 \times 3 = 21$
• $5 \times 7 = 35$, so $7 \times 5 = 35$
• $9 \times 7 = 63$, so $7 \times 9 = 63$

The multiplication facts for 7 lead to the inverse division facts.

• $7 \times 3 = 21$, so $21 \div 7 = 3$
• $7 \times 4 = 28$, so $28 \div 7 = 4$
• $7 \times 8 = 56$, so $56 \div 7 = 8$

Write the 7 times table.

Draw a ring around all the facts you know from commutative facts for other tables.

Underline the new fact.

Guided practice

Use these numbers to write two multiplication and two division facts.

4, 7, 28 | $4 \times 7 = 28$, $7 \times 4 = 28$, $28 \div 7 = 4$, $28 \div 4 = 7$

Lesson 1: **Multiples**

- Understand and identify multiples

Key words
- multiply
- product
- multiple
- integer

Let's learn

A **multiple** is the result of multiplying a number by an integer.

×	1	2	3	4	5	6	7	8	9	10
1	1	2	3	4	5	6	7	8	9	10
2	2	4	6	8	10	12	14	16	18	20
3	3	6	9	12	15	18	21	24	27	30
4	4	8	12	16	20	24	28	32	36	40
5	5	10	15	20	25	30	35	40	45	50
6	6	12	18	24	30	36	42	48	54	60
7	7	14	21	28	35	42	49	56	63	70
8	8	16	24	32	40	48	56	64	72	80
9	9	18	27	36	45	54	63	72	81	90
10	10	20	30	40	50	60	70	80	90	100

This row shows the **multiples of 1**.

This row shows the **multiples of 2**.

This row shows the **multiples of 3**.

From the table I can see that 12 is a multiple of 2 and 6, and of 3 and 4. I also know that 12 is a multiple of 1 and 12.

Look at the table.
Write three numbers that are multiples of 1, 2, 3 and 4.

Guided practice

How can you prove that 20 is a multiple of 2, 4, 5 and 10?

20 comes in the 2, 4, 5 and 10 times tables. 2 × 10 = 20, 4 × 5 = 20, 5 × 4 = 20 and 10 × 2 = 20.

Lesson 2: **Factors**

> • Understand and identify factors

Number

Let's learn

This multiplication table shows that 18 is in the 2, 3, 6 and 9 times tables.
So 2, 3, 6 and 9 are all factors of 18.
1 and 18 are also factors of 18.

We can use multiplication and division facts to show this.
$2 \times 9 = 18$ and $18 \div 2 = 9$.
So 2 and 9 are both factors of 18.

A **factor** is a whole number that divides exactly into another number.

×	1	2	3	4	5	6	7	8	9	10
1	1	2	3	4	5	6	7	8	9	10
2	2	4	6	8	10	12	14	16	18	20
3	3	6	9	12	15	18	21	24	27	30
4	4	8	12	16	20	24	28	32	36	40
5	5	10	15	20	25	30	35	40	45	50
6	6	12	18	24	30	36	42	48	54	60
7	7	14	21	28	35	42	49	56	63	70
8	8	16	24	32	40	48	56	64	72	80
9	9	18	27	36	45	54	63	72	81	90
10	10	20	30	40	50	60	70	80	90	100

Look at the table. Write three multiples and their factors.

Guided practice

How can you prove that 3 is a factor of 6, 12, 24 and 30?

The numbers are all products in the 3 times table so 3 must be a factor of all the numbers.
$6 \div 3 = 2$, $12 \div 3 = 4$, $24 \div 3 = 8$ and $30 \div 3 = 10$.

Number

Lesson 3: **Multiples and factors**

Key words
- **multiply**
- **product**
- **divide**
- **quotient**
- **multiple**
- **factor**

- Understand the relationship between multiples and factors

Let's learn

Multiples are made when **factors** are multiplied together.

Factors are the result of dividing a number by another number exactly.

We can say:

factor × factor = multiple   multiple ÷ factor = factor

Look at this example:

$$8 \quad \times \quad 4 \quad = \quad 32$$

factor   factor   multiple

$$32 \quad \div \quad 4 \quad = \quad 8$$

multiple   factor   factor

What two numbers can be multiplied together to make 28?

Write the multiplication and division statements to show how multiples and factors are linked together.

Guided practice

Fill in the boxes to make these calculations correct.

a $\boxed{3} \times 9 = 27$

$27 \div \boxed{9} = 3$

b $\boxed{9} \times 4 = 36$

$36 \div \boxed{4} = 9$

Lesson 4: **Tests of divisibility**

- Understand tests of divisibility by 2, 5, 10, 25, 50 and 100

Key words
- **multiply**
- **product**
- **divide**
- **quotient**
- **multiple**
- **factor**

Let's learn

Multiples of 2 all have an even ones digit.

This means that all even numbers can be divided by 2.

Multiples of 5 have 5 or 0 as the ones digit.

This means that all numbers with 5 or 0 in the ones position can be divided by 5.

Multiples of 10 have 0 as the ones digit.

This means that all numbers with 0 in the ones position can be divided by 10.

2	5	10
4	10	20
6	15	30
8	20	40
10	25	50
12	30	60
14	35	70
16	40	80
18	45	90
20	50	100

This table shows the multiples of 2, 5 and 10 up to 20, 50 and 100.

This table shows the first 10 multiples of 25, 50 and 100.

25	50	75	100	125	150	175	200	225	250
50	100	150	200	250	300	350	400	450	500
100	200	300	400	500	600	700	800	900	1000

25 is a factor of 50 and 100.

50 is a factor of 100.

Look at the row beginning with 25. Are all these numbers divisible by 2, 5 and 10? Discuss this with your partner.

Look at the rows beginning with 50 and 100. Are all these numbers divisible by 2, 5 and 10? Discuss this with your partner.

Guided practice

Draw a ring around the odd one out.

(120)   150   175   250   325

Why is it the odd one out?

It is the only number that isn't a multiple of 25.

Number

Number

Lesson 1: **Grouping numbers in different ways**

- Group numbers in different ways to make multiplication simpler

Let's learn

Using **factors** makes multiplication easy.

Regroup the multiplicand into factors. Then choose how to multiply. Don't forget to estimate the product first.

16×3 will be between 40 and 60.

$16 \times 3 = 8 \times 2 \times 3$ or $16 \times 3 = 4 \times 4 \times 3$

$= \quad 24 \times 2$ $= \quad 12 \times 4$

For $8 \times 2 \times 3$, we can multiply 8 by 3 and then double. $8 \times 3 = 24$ Double 24 is 48.

For $4 \times 4 \times 3$, we can multiply 4 by 3 and then double twice. $4 \times 3 = 12$ Double 12 is 24. Double 24 is 48.

This is called the **associative property** of multiplication. It doesn't matter what order we multiply numbers in, the product will always be the same.

How can you use factors to work out the product of 32 and 3?

4 Share ideas with your partner.

Which is the simplest way to find the product?

Guided practice

How can you use this calculation to answer 36×5?

$9 \times 4 \times 5 =$

$9 \times 5 = 45$
$45 \times 2 = 90$
$90 \times 2 = 180$

Lesson 2: **Multiplying tens**

- Multiply tens numbers by 1-digit numbers

Number

Let's learn

$20 \times 5 = 100$	$30 \times 5 = 150$

$40 \times 5 = 200$	$50 \times 5 = 250$

It's easy to multiply tens numbers!

We can work out 30×5, by working out $3 \times 5 = 15$ and then $15 \times 10 = 150$.

Just look at the tens digit of the number, multiply it by the multiplier and then multiply by 10.

This number line shows the 30 times table.

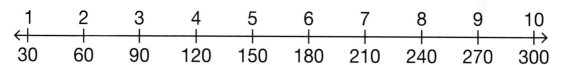

1	2	3	4	5	6	7	8	9	10
30	60	90	120	150	180	210	240	270	300

We can work out the 30 times table by multiplying each of the numbers 1 to 10 by 3.

Then multiply the product by 10.

Draw a number line to show the 90 times table.

What do you notice about the number line for 30 and the one for 90?

Guided practice

How can you multiply 80 by 8?

$8 \times 8 = 64$ and $64 \times 10 = 640$.

So, $80 \times 8 = 640$.

Number

Lesson 3: **Multiplying hundreds**

Key words
• multiply
• multiplicand
• multiplier
• product
• hundreds
• one-digit number

• Multiply hundreds numbers by 1-digit numbers

Let's learn

| 200 × 5 = 1000 | 300 × 5 = 1500 |

| 400 × 5 = 2000 | 500 × 5 = 2500 |

It's easy to multiply hundreds numbers!

Just look at the hundreds digit of the number, multiply it by the multiplier and then multiply by 10 and 10 again.

Multiplying by 10 and 10 again is the same as multiplying by 100.

We can work out 300 × 5 by working out 3 × 5 = 15 and then 15 × 10 × 10 = 1500.

This number line shows the 400 times table.

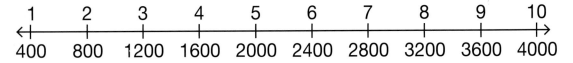

| 1 | 2 | 3 | 4 | 5 | 6 | 7 | 8 | 9 | 10 |
| 400 | 800 | 1200 | 1600 | 2000 | 2400 | 2800 | 3200 | 3600 | 4000 |

We can work out the 400 times table by multiplying each of the numbers 1 to 10 by 4.

Then multiply the product by 100.

Draw a number line to show the 800 times table.

What do you notice about the number line for 400 and the one for 800?

Guided practice

How can you multiply 400 by 6?

4 × 6 = 24 and 24 × 100 = 2400.
So, 400 × 6 = 2400.

Lesson 4: **Multiplying 2-digit numbers and ones (1)**

Key words
- estimate
- multiply
- product
- partition
- exchange

- Estimate and multiply 2-digit numbers by 1-digit numbers

Let's learn

This is the **array** for calculating 43 × 4.

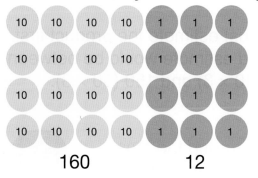

160   12

Don't forget to estimate the product first. A good estimate would be between 160 (40 × 4) and 200 (50 × 4).

Exchange 10 tens for 1 hundred and 10 ones for 1 ten to give 172.

| 100 | 10 | 10 | 10 | 10 | 10 | 10 | 10 | 1 | 1 |

This is the **grid method** for 43 × 4.

×	40	3
4	160	12

160
+ 12
—
172

This is the **partitioning method** for 43 × 4.

43 × 4 =

40 × 4   3 × 4

160   +   12   =   172

With your partner, use place value counters to show 43 × 5.
Estimate the product first. Find the product.
Now use the grid and partitioning methods for the same calculation.

Guided practice

Use partitioning to answer the calculation.

54 × 6 = $\boxed{324}$

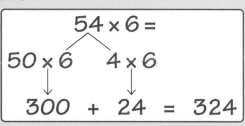

54 × 6 =

50 × 6   4 × 6

300   +   24   =   324

Lesson 1: **Multiplying 2-digit numbers and ones (2)**

Key words
* multiplicand
* multiplied by
* multiplier
* product

- Estimate and multiply 2-digit numbers by 1-digit numbers

Let's learn

We can use this array to work out 44 × 3.

10 10 10 10 1 1 1 1
10 10 10 10 1 1 1 1
10 10 10 10 1 1 1 1

We must estimate the product first. A good estimate would be between 120 (40 × 3) and 150 (50 × 3).

Group the ones and exchange 10 ones for 1 ten.

Then group the tens and exchange 10 tens for 1 hundred.

100 10 10 10 1 1   44 × 3 = 132

We know the grid method and the partitioning method to calculate 44 × 3.
This new method is the **expanded written method**.

```
      4 4
  ×     3
      1 2      4 × 3
  + 1 2 0     40 × 3
    1 3 2
```

👥 With your partner, use place value counters to show 28 × 3.
Estimate the product first. Find the product.
Now work out the answer using the expanded written method.

Guided practice

Use the expanded written method to answer the calculation.

48 × 3 = ⎡ 144 ⎤

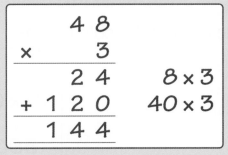

```
      4 8
  ×     3
      2 4      8 × 3
  + 1 2 0     40 × 3
    1 4 4
```

Lesson 2: **Multiplying 3-digit numbers and ones (1)**

• Estimate and multiply 3-digit numbers by 1-digit numbers

Let's learn

We can use this array to work out 243 × 3.

We group and exchange the different values as before.

×	**200**	**40**	**3**
	100 100	10 10 10 10	1 1 1
3	100 100	10 10 10 10	1 1 1
	100 100	10 10 10 10	1 1 1

We must estimate the product first. A good estimate would be between 600 (200 × 3) and 900 (300 × 3).

We can also use the grid method.

×	200	40	3
3	600	120	9

$$\begin{array}{r} 600 \\ 120 \\ + \quad 9 \\ \hline 729 \end{array}$$

• With your partner, use place value counters to show 238 × 6.

Estimate the product first. Find the product.

Now show the same calculation using the grid method.

Guided practice

Use the grid method to answer the calculation.

135 × 6 = 810

×	100	30	5
6	600	180	30

$$\begin{array}{r} 600 \\ 180 \\ + \quad 30 \\ \hline 810 \\ {\scriptstyle 1} \end{array}$$

Lesson 3: **Multiplying 3-digit numbers and ones (2)**

* Estimate and multiply 3-digit numbers by 1-digit numbers

Key words
* multiplicand
* multiplied by
* multiplier
* product

Let's learn

We can use arrays, the grid method, partitioning and the expanded written method to multiply 3-digit numbers by 1-digit numbers.

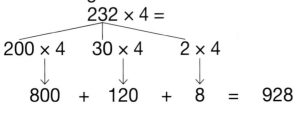

×	200	30	2

The grid method

×	200	30	2
4	800	120	8

$$\begin{array}{r} 800 \\ 120 \\ + \quad 8 \\ \hline 928 \end{array}$$

Partitioning

$$232 \times 4 =$$

$200 \times 4 \qquad 30 \times 4 \qquad 2 \times 4$

$$800 \; + \; 120 \; + \; 8 \; = \; 928$$

Expanded written method

$$\begin{array}{r} 2\ 3\ 2 \\ \times \qquad 4 \\ \hline 8 \\ 1\ 2\ 0 \\ + \ 8\ 0\ 0 \\ \hline 9\ 2\ 8 \end{array}$$

2×4
30×4
200×4

👥 Discuss each of these methods with your partner.

◁4 Describe how each one works.

Which do you prefer?

Try to convince your partner that it is the best one to use!

Guided practice

Use the expanded written method to answer the calculation.

$239 \times 7 = \boxed{1673}$

$$\begin{array}{r} 2\ 3\ 9 \\ \times \qquad 7 \\ \hline 6\ 3 \\ 2\ 1\ 0 \\ + \ 1\ 4\ 0\ 0 \\ \hline 1\ 6\ 7\ 3 \end{array}$$

9×7
30×7
200×7

Lesson 4: **Multiplying 3-digit numbers and ones (3)**

Number

• Estimate and multiply 3-digit numbers by 1-digit numbers

Let's learn

Tilly has 145 coins. Omari has 6 times as many. How many coins does Omari have?

In this problem we use the words 'times as many'. This phrase tells us that we need to multiply.

A good way to find the product of 145 and 6 is to use the expanded written method. Let's estimate the product first. A good estimate would be 900 (150 × 6).

```
    1 4 5
  ×     6
      3 0      5 × 6
    2 4 0     40 × 6
  + 6 0 0    100 × 6
    8 7 0
```

With your partner, make up a problem for 234 × 7.

Estimate first, then find the product. Use your preferred method. Your partner checks your answer using a different method.

Guided practice

Yukesh has 128 stamps. Sheba has 3 times as many. How many stamps does Sheba have? Use partitioning to find out.

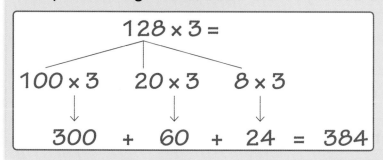

128 × 3 =

100 × 3   20 × 3   8 × 3

300 + 60 + 24 = 384

Lesson 1: **Dividing 2-digit numbers by 2 and 4**

- Estimate and divide 2-digit numbers by 2 and 4

Key words
- dividend
- divided by
- divisor
- quotient
- remainder

Let's learn

$76 \div 2 =$

To divide even numbers by 2, we can use partitioning and halving.

A good estimate would be between 30 and 40.

Dividing an odd number by 2 always gives a remainder of 1.

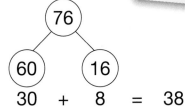

$$30 \; + \; 8 \; = \; 38$$

$96 \div 4 =$

To divide a number by 4, partition and halve each part twice.

When we divide by 4, the remainder can be 1, 2 or 3.

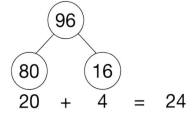

$$20 \; + \; 4 \; = \; 24$$

Think of three numbers that will have a remainder of 1 when you divide them by 2.

Divide them by 2 to see if you are correct.

Guided practice

Think of a number that will have a remainder of 3 when divided by 4.

Divide your number by 4 to see if you are correct.

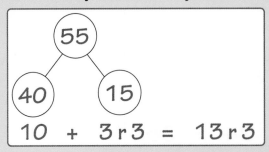

$$10 \; + \; 3 \, r \, 3 \; = \; 13 \, r \, 3$$

Lesson 2: **Dividing 2-digit numbers by 5**

Key words
- dividend
- divided by
- divisor
- quotient
- remainder

Number

- Estimate and divide 2-digit numbers by 5

Let's learn

A useful mental strategy when dividing a multiple of 10 by 5 is to first divide by 10 and then double.

$90 \div 5 =$

$90 \div 10 = 9$

Double $9 = 18$

So, $90 \div 5 = 18$

To divide multiples of 5 that are not multiples of 10, partition and use known times tables facts.

$65 \div 5 =$

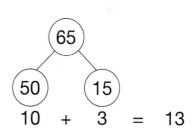

65

50   15

10 + 3 = 13

Any number that is not a multiple of 5 or 10 will have a remainder when it is divided by 5.

$73 \div 5 =$

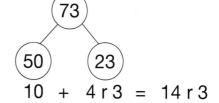

73

50   23

10 + 4 r 3 = 14 r 3

Think of three numbers that will have a remainder of 1 when divided by 5. Now think of three numbers that will have a remainder of 2 when divided by 5.

Guided practice

Use partitioning to answer: $85 \div 5 =$

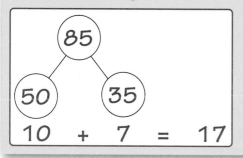

85

50   35

10 + 7 = 17

Lesson 3: **Dividing 2-digit numbers by 3 and 6**

* Estimate and divide 2-digit numbers by 3 and 6

Key words
* dividend
* divided by
* divisor
* quotient
* remainder

Let's learn

$74 \div 3 =$

To divide by 3, we can take away groups of 3.

Or we can use partitioning and known times tables facts.

We should estimate the quotient first. A good estimate would be between 20 and 30.

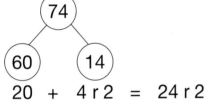

74
60   14

$20 \ + \ 4 \, r \, 2 \ = \ 24 \, r \, 2$

$74 \div 6 =$

When we divide the same number by 6, the quotient will be half what it is when we divide by 3.

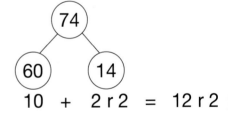

74
60   14

$10 \ + \ 2 \, r \, 2 \ = \ 12 \, r \, 2$

With your partner, use place value counters to make 84.

Partition 84 into 60 and 24.

Use known times tables facts to divide each part by 3.

Then repeat but divide by 6.

What are the two quotients?

Guided practice

Use partitioning to answer: $68 \div 3 =$

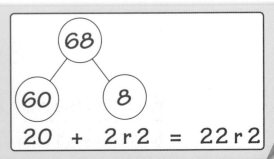

68
60   8

$20 \ + \ 2 \, r \, 2 \ = \ 22 \, r \, 2$

Number

Lesson 4: **Dividing 2-digit numbers by 7, 8 and 9**

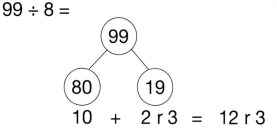
• Estimate and divide 2-digit numbers by 7, 8 and 9

Let's learn

We can use known times tables facts to divide numbers by 7, 8 or 9.

If a number has more than 10 groups of the divisor, partition these first.

$99 \div 7 =$

```
        99
       /  \
     70    29
   10  +  4 r 1  =  14 r 1
```

$99 \div 8 =$

```
        99
       /  \
     80    19
   10  +  2 r 3  =  12 r 3
```

$99 \div 9 =$

```
        99
       /  \
     90    9
   10  +  1  =  11
```

Knowing off by heart all the times tables facts up to 10×10 makes division easier.

Talk to your partner about remainders.

What remainders can you have when dividing by 7?

Write an example of a number that will have a remainder of 5.

What remainders can you have when dividing by 8?

Write an example of a number that will have a remainder of 6.

What remainders can you have when dividing by 9?

Write an example of a number that will have a remainder of 7.

Guided practice

Divide each number by 9.
Don't forget the remainder.

a 78 $\boxed{8 \text{ r } 6}$

b 59 $\boxed{6 \text{ r } 5}$

c 65 $\boxed{7 \text{ r } 2}$

Lesson 1: **Working towards a written method**

Key words
- **dividend**
- **divided by**
- **divisor**
- **quotient**
- **remainder**

- Estimate and divide 2-digit numbers by 1-digit numbers

Number

Let's learn

To divide, we take away groups of the divisor.

To solve 54 ÷ 4 = we must find how many groups of 4 there are in 54.

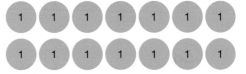

We should estimate the quotient first. A good estimate would be between 10 and 20.

Make one group of 4 tens…

and exchange the remaining ten for 10 ones to make 14 ones altogether.

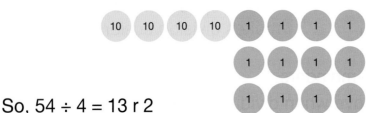

Now there is one group of 4 tens… and three groups of 4 ones.

So, 54 ÷ 4 = 13 r 2

👥 Talk to your partner about how you can check that the answer of
4 13 remainder 2 is correct.

Try out your ideas.

Guided practice

Use place value counters to calculate 42 ÷ 3 =.
Draw a diagram and explain what you did.

42 ÷ 3 = [14]

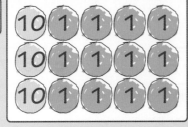

I made one group of 3 tens.
I exchanged the leftover
ten for 10 ones.
I made 4 groups of 3 ones.

Lesson 2: **Written method of division (1)**

Number

* Estimate and divide 2-digit numbers by 1-digit numbers using a written method

Let's learn

To work out $73 \div 3 =$ we must find out how many groups of 3 we can make out of 73.
First we estimate. The quotient will be between 20 and 30.

10s	1s
× × × × × × ×	× × ×

$$3\overline{)7\ 3}$$

We can make 2 groups of 3 tens.
We need to exchange the remaining ten for 10 ones.

10s	1s
(× × ×)	× × ×
(× × ×)	
×	

$$\begin{array}{r} 2 \\ 3\overline{)7\ 3} \end{array}$$

We then regroup them with the other ones.

10s	1s
(× × ×)	× × × × × × ×
(× × ×)	× × × × × ×

$$\begin{array}{r} 2 \\ 3\overline{)7\ 3} \\ -\ 6\ 0 \\ \hline 1\ 3 \end{array}$$

Now we make groups of 3 ones.
We have made 4 groups of 3 ones.
There is 1 remaining.
So, $73 \div 3 = 24$ r 1

10s	1s
(× × ×)	(× × ×)
(× × ×)	(× × ×)
	(× × ×)
	(× × ×)
	×

$$\begin{array}{r} 2\ 4\ r\ 1 \\ 3\overline{)7\ 3} \\ -\ 6\ 0 \\ \hline 1\ 3 \\ -\ 1\ 2 \\ \hline 1 \end{array}$$

* Think of a 2-digit number that has 9 tens. Divide it by 4.
Estimate the quotient first.
Use place value counters and draw visual representations.
Record using the expanded written method.

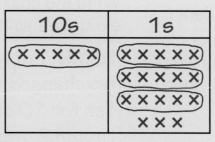

Number

Lesson 3: **Written method of division (2)**

• Estimate and divide 2-digit numbers by 1-digit numbers using a written method

Let's learn

To work out $87 \div 7 =$ we must find out how many groups of 7 we can make out of 87.

First we must estimate. A good estimate would be between 10 and 20.

10s	1s
× × × × × × × ×	× × × × × × ×

$$7\overline{)8\ 7}$$

We can make 1 group of 7 tens. We need to exchange the remaining ten for 10 ones.

10s	1s
(× × × × × × ×)×	× × × × × × ×

$$\begin{array}{r} 1 \\ 7\overline{)8\ 7} \end{array}$$

We then regroup them with the other ones.

10s	1s
(× × × × × × ×)	× × × × × × ×
	× × × × ×
	× × × × ×

$$\begin{array}{r} 1 \\ 7\overline{)8\ 7} \\ -\ 7\ 0 \\ \hline 1\ 7 \end{array}$$

Now we make groups of 7 ones. We have made 2 groups of 7 ones. There are 3 remaining.

So, $87 \div 7 = 12$ r 3

10s	1s
(× × × × × × ×)	(× × × × × × ×)
	(× × × × ×)
	(× ×)× × ×

$$\begin{array}{r} 1\ 2\ \text{r}\ 3 \\ 7\overline{)8\ 7} \\ -\ 7\ 0 \\ \hline 1\ 7 \\ -\ 1\ 4 \\ \hline 3 \end{array}$$

👥 Think of a 2-digit number that is greater than 80. Divide it by 7.

Estimate the quotient first. Use place value counters. Use the expanded written method to record what you do.

Guided practice

Explain how to divide 96 by 8. Write the calculation using the expanded written method.

$$\begin{array}{r} 1\ 2 \\ 8\overline{)9\ 6} \\ -\ 8\ 0 \\ \hline 1\ 6 \\ -\ 1\ 6 \\ \hline 0 \end{array}$$

*I made 1 group of 8 tens.
I exchanged the left over ten for 10 ones.
I made 2 groups of 8 ones.*

Lesson 4: **Using division to solve problems**

• Estimate and divide 2-digit numbers by
1-digit numbers to solve problems

Key words
• **dividend**
• **divided by**
• **divisor**
• **quotient**
• **remainder**

Let's learn

One of the reasons that we divide is to solve problems.
Look at this problem.

There are 96 pencils in the classroom.

The teacher puts 8 pencils in each pot.

How many pots does she fill?

This means that we need to imagine taking 1 group of
8 pencils and putting them in a pot. We keep doing this
until all the pencils have been put in pots.

This is the same as dividing 96 by 8.

With your partner, work out the number of pots the
teacher has filled.

One of you do this using a partitioning method.

One of you use the expanded written method.

Which method do you prefer?

Why?

Guided practice

Mandy and Leanne pick 65 flowers.

They put three flowers in each vase.

How many vases do they use? $\boxed{21}$

How many flowers are left? $\boxed{2}$

How did you work that out?

I partitioned 65 into 60 and 5. I divided 60 by 3 to give 20.
I divided 5 by 3 to give 1 remainder 2.

Number

Lesson 1: **Understanding place value (1)**

Key words
- **thousands**
- **hundreds**
- **tens**
- **ones**
- **position**
- **multiply**
- **add**

- Understand the value of each digit in a 4-digit number

Let's learn

Look at this place value chart.

> The digit 4 is in the 100s position.

> The digit 8 is in the 10s position.

> The digit 2 is in the 1000s position.

1000s	100s	10s	1s
2	4	8	3

> The digit 3 is in the 1s position.

To find the value of each digit, look at its position in the place value chart.

The 2 represents 2 thousands.   The 4 represents 4 hundreds.

$2 \times 1000 = 2000$ $4 \times 100 = 400$

The 8 represents 8 tens.   The 3 represents 3 ones.

$8 \times 10 = 80$ $3 \times 1 = 3$

To find the value of the whole number, add the numbers together.

$$2000 + 400 + 80 + 3 = 2483$$

We can represent 2314 on an abacus and using place value cards.

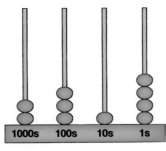

2000 > 300 > 10 > 4

1000s 100s 10s 1s

Talk to your partner about the number 6534.

What positions are the digits in?

What are their values?

What do we do to make the whole number?

Guided practice

What is the value of each digit?

$3592 = \boxed{3000} + \boxed{500} + \boxed{90} + \boxed{2}$

Lesson 2: **Understanding place value (2)**

• Understand the value of each digit in a 5-digit number

Key words
• tens of thousands
• thousands
• hundreds
• tens
• ones
• position
• multiply
• add
• place holder

Let's learn

10 000s	1000s	100s	10s	1s
4	6	3	7	2

This is a 5-digit number.
There are two thousands numbers.
The first is 4 ten thousands.
The second is 6 thousands.

The whole number is forty-six thousand, three hundred and seventy-two.
Multiply each digit by the number at the top of the column.
Then add all the values together.
40 000 + 6000 + 300 + 70 + 2

10 000s	1000s	100s	10s	1s
4	6	0	7	2

This number has no hundreds so there is a zero in the 100s position. The zero is a place holder.

Write some 5-digit numbers that have a zero in the 100s position.
Now write some numbers that have a zero in the 1000s position.
Explain to your partner what the zero does.

Guided practice
What is the value of each digit?

35 914 = [30 000] + [5000] + [900] + [10] + [4]

Number

Lesson 3: **Regrouping**

- Regroup 4-digit numbers in different ways

Let's learn

This is 1243 in groups of thousands, hundreds, tens and ones.

(1000) (100) (100) (10) (10) (10) (10) (1) (1) (1)

$1000 + 200 + 40 + 3 = 1243$

We can **regroup** 1243 as 1243 ones.

We can also regroup 1243 as 124 tens and 3 ones.

Or we can regroup 1243 as 12 hundreds, 4 tens and 3 ones.

Here are some other ways to regroup 1243.

- 12 hundreds and 43 ones
- 1 thousand, 22 tens and 23 ones
- 11 hundreds and 14 tens and 3 ones

Regroup this number in as many ways as you can.

1

(1000) (100) (100) (100) (10) (10) (1) (1)

Start with ones.

Then look at tens and ones.

Then hundreds, tens and ones.

Then think of some more ways.

Guided practice

Sabrina has made the number 2398.

She thinks the only way to group her number is 2000, 300, 90 and 8.

Sabrina is incorrect.

Explain why, with two examples to prove your thinking.

> There are lots of ways to group numbers. Sabrina could have grouped her number as 239 tens and 8 ones. She could have grouped it as 2398 ones.

Lesson 4: **Comparing and ordering numbers**

• Compare and order numbers, including negative numbers

Let's learn

To compare and order 4-digit numbers, start by looking at the thousands digits.

6456

6456 has 6 thousands.
6456 is greater than 2356.
6456 > 2356

2356

2356 has 2 thousands.
2356 is less than 6456.
2356 < 6456

Now look at these numbers. The thousands digits are the same, so look at the hundreds digits.

4436

4436 has 4 hundreds.
4436 is less than 4856.
4436 < 4856

4856

4856 has 8 hundreds.
4856 is greater than 4436.
4856 > 4436

We use the same idea for ordering.

7165   **7135**   **7185**

The thousands digits are the same and the hundreds digits are the same, so look at the tens digits.

7135   **7165**   **7185**

Write six 4-digit numbers on separate slips of paper.

Choose pairs of numbers and compare the numbers, smallest first.

Then take groups of four numbers and order them, smallest first.

Guided practice

Write > or < between each pair of numbers.

a 5436 > 2456   **b** 1243 < 1271

Lesson 1: **Multiplying and dividing by 10**

* Use place value to multiply and divide whole numbers by 10

Let's learn

When we **multiply by 10**, the value of each digit becomes **10 times greater** and the digits move **one place value to the left**. We include a 0 in the ones place to act as a place holder.

$$5687 \times 10 = 56\,870$$

10 000s	1000s	100s	10s	1s
	5	6	8	7
5	6	8	7	0

When we **divide by 10**, the value of each digit becomes **10 times smaller** and the digits move **one place value to the right**.

$$4190 \div 10 = 419$$

10 000s	1000s	100s	10s	1s
	4	1	9	0
		4	1	9

Make a list of five 3-digit numbers.

Multiply each number by 10. Write the product.

Explain to your partner what has happened to your original number.

Make a list of five 4-digit multiples of 10, such as 4380.

Divide each number by 10. Write the quotient.

Explain to your partner what has happened to your original number.

Guided practice

Multiply each number by 10.   Divide each number by 10.

a 371 ┃ 3710 ┃   **a** 2450 ┃ 245 ┃

b 8746 ┃ 87 460 ┃   **b** 56 780 ┃ 5678 ┃

Lesson 2: **Multiplying and dividing by 100**

Key words
* **multiply**
* **product**
* **divide**
* **quotient**
* **increase**
* **decrease**
* **place holder**

• Use place value to multiply and divide whole numbers by 100

Let's learn

When we **multiply by 100**, the value of each digit becomes **100 times greater** and the digits move **two place values to the left**. We include a 0 in the tens and ones places to act as place holders.

$$328 \times 100 = 32\,800$$

10 000s	1000s	100s	10s	1s
		3	2	8
3	2	8	0	0

When we **divide by 100**, the value of each digit becomes **100 times smaller** and the digits move **two place values to the right**.

$$57\,400 \div 100 = 574$$

10 000s	1000s	100s	10s	1s
5	7	4	0	0
		5	7	4

Make a list of five 3-digit numbers.

Multiply each number by 100. Write the product.

Explain to your partner what has happened to your original number.

Make a list of five 4-digit multiples of 100, such as 2700.

Divide each number by 100. Write the quotient.

Explain to your partner what has happened to your original number.

Guided practice

Multiply each number by 100.   Divide each number by 100.

a 378 | 37 800 | **a** 78 600 | 786 |

b 956 | 95 600 | **b** 45 700 | 457 |

Number

Lesson 3: **Rounding to the nearest 10, 100 and 1000**

• Round numbers to the nearest 10, 100 and 1000

Let's learn

73 + 98 =

We can estimate the total by rounding both numbers to the nearest 10.

This gives 70 + 100 = 170.

An estimate is not accurate, but gives an approximate answer.

This signpost reminds us when to round up or down.

Round down	Round up
0 1 2 3 4	5 6 7 8 9

5438 rounded to the nearest 10 is 5440 because 8 in the ones position rounds up.

5438 rounded to the nearest 100 is 5400 because the 3 in the tens position rounds down.

5438 rounded to the nearest 1000 is 5000 because the 4 in the hundreds position rounds down.

With your partner write three numbers that would be rounded to 5490. Now write three numbers that would be rounded to 2400.

Guided practice

Round these numbers to the nearest 100.

a 279 ☐ 300

b 7698 ☐ 7700

Lesson 4: **Rounding to the nearest 10 000 and 100 000**

Key words
- rounding
- round up
- round down
- thousands
- ten thousands
- hundred thousands
- millions

- Round numbers to the nearest 10 000 and 100 000

Let's learn

1 000 000s	100 000s	10 000s	1000s	100s	10s	1s
1	4	3	8	7	3	9

This number is 1 million, 438 thousand, 7 hundred and 39.

We can round this number in the same way as other numbers.

To round to the nearest 10 000, we look at the digit in the 1000s position. It is 8, so we round up to give 1 440 000.

To round to the nearest 100 000, we look at the digit in the 10 000s position. It is 3, so we round down to give 1 400 000.

With your partner, make a list of three 7-digit numbers.

Each digit in each of your numbers must be different, for example, 1 846 739

Round each number to the nearest 10 000 and 100 000.

Record your answers.

Explain to each other how you know which digit to round, and how to round it.

Guided practice
Round each number to the nearest 10 000.

a 64 492 $\boxed{60\,000}$

b 56 235 $\boxed{60\,000}$

c 245 892 $\boxed{250\,000}$

d 715 328 $\boxed{720\,000}$

Number

Lesson 1: **Understanding fractions**

- Understand that the larger the denominator, the smaller the fraction

Let's learn

$$\frac{1}{2}$$

The numerator says how many of the parts there are in the fraction.

The denominator says how many equal parts there are in the whole.

The whole has been divided into 2 equal parts. Each part is $\frac{1}{2}$.

| $\frac{1}{2}$ | $\frac{1}{2}$ |

The whole has been divided into 3 equal parts. Each part is $\frac{1}{3}$.

| $\frac{1}{3}$ | $\frac{1}{3}$ | $\frac{1}{3}$ |

The whole has been divided into 5 equal parts. Each part is $\frac{1}{5}$.

| $\frac{1}{5}$ | $\frac{1}{5}$ | $\frac{1}{5}$ | $\frac{1}{5}$ | $\frac{1}{5}$ |

The whole has been divided into 8 equal parts. Each part is $\frac{1}{8}$.

| $\frac{1}{8}$ | $\frac{1}{8}$ | $\frac{1}{8}$ | $\frac{1}{8}$ | $\frac{1}{8}$ | $\frac{1}{8}$ | $\frac{1}{8}$ | $\frac{1}{8}$ |

Which is the smallest fraction?
Which is the greatest fraction?
Explain how you know.

$\frac{3}{7}$ $\frac{3}{5}$ $\frac{3}{4}$

Guided practice

Ling eats a slice of this pizza.

What fraction does she eat? $\boxed{\frac{1}{3}}$

What fraction is left? $\boxed{\frac{2}{3}}$

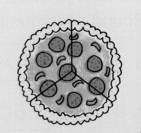

Lesson 2: **Making one whole**

- Understand that fractions can be combined to make one whole

Let's learn

1							
$\frac{1}{2}$				$\frac{1}{2}$			
$\frac{1}{4}$		$\frac{1}{4}$		$\frac{1}{4}$		$\frac{1}{4}$	
$\frac{1}{8}$	$\frac{1}{8}$	$\frac{1}{8}$	$\frac{1}{8}$	$\frac{1}{8}$	$\frac{1}{8}$	$\frac{1}{8}$	$\frac{1}{8}$

1											
$\frac{1}{3}$				$\frac{1}{3}$				$\frac{1}{3}$			
$\frac{1}{6}$		$\frac{1}{6}$		$\frac{1}{6}$		$\frac{1}{6}$		$\frac{1}{6}$		$\frac{1}{6}$	
$\frac{1}{12}$	$\frac{1}{12}$	$\frac{1}{12}$	$\frac{1}{12}$	$\frac{1}{12}$	$\frac{1}{12}$	$\frac{1}{12}$	$\frac{1}{12}$	$\frac{1}{12}$	$\frac{1}{12}$	$\frac{1}{12}$	$\frac{1}{12}$

We can combine $\frac{1}{4}$ and $\frac{6}{8}$ to make 1 whole, because $\frac{1}{4}$ is equal to $\frac{2}{8}$. $2 + 6 = 8$, therefore $\frac{2}{8}$ and $\frac{6}{8} = \frac{8}{8} = 1$ whole.

We can combine $\frac{1}{3}$ and $\frac{8}{12}$ to make 1 whole, because $\frac{1}{3}$ is equivalent to $\frac{4}{12}$. $4 + 8 = 12$, so $\frac{4}{12}$ and $\frac{8}{12} = \frac{12}{12} = 1$.

Combine three different fractions together to make one whole. Use the fraction wall to help you.

Prove to your partner that you have made one whole.

Draw a bar model to show what you have done.

Guided practice

Samson writes two fractions: $\frac{4}{6}$ $\frac{1}{3}$

a What is the same about the fractions?

> They have the same value.

b What is different?

> They are made from different numbers.

c If Samson puts the two fractions together what will he make?

> One whole

Lesson 3: **Equivalent fractions**

- Recognise that two proper fractions can be equivalent in value

Key words
- fraction
- unit fraction
- non-unit fraction
- whole
- part
- equal
- equivalent
- denominator
- numerator

Let's learn

$\frac{1}{5}$		$\frac{1}{5}$		$\frac{1}{5}$		$\frac{1}{5}$		$\frac{1}{5}$	
$\frac{1}{10}$	$\frac{1}{10}$	$\frac{1}{10}$	$\frac{1}{10}$	$\frac{1}{10}$	$\frac{1}{10}$	$\frac{1}{10}$	$\frac{1}{10}$	$\frac{1}{10}$	$\frac{1}{10}$

$\frac{1}{15}$	$\frac{1}{15}$	$\frac{1}{15}$	$\frac{1}{15}$	$\frac{1}{15}$	$\frac{1}{15}$	$\frac{1}{15}$	$\frac{1}{15}$	$\frac{1}{15}$	$\frac{1}{15}$	$\frac{1}{15}$	$\frac{1}{15}$	$\frac{1}{15}$	$\frac{1}{15}$	$\frac{1}{15}$

From the fraction wall, $\frac{1}{5} = \frac{2}{10} = \frac{3}{15}$

To find **equivalent fractions**, multiply or divide the numerator and the denominator of the first fraction by the same number.

If I multiply the numerator and denominator of $\frac{1}{5}$ by 2, I will get the equivalent fraction of $\frac{2}{10}$.
If I multiply them both by 3 I will get $\frac{3}{15}$.

It is the same for non-unit fractions.
If I multiply the numerator and denominator of $\frac{2}{5}$ by 2 I get $\frac{4}{10}$.

Use the multiplication table resource sheet to write three fractions that are equivalent to $\frac{4}{5}$.
Talk to your partner about what you would multiply the numerator and denominator by to get your equivalent fractions.

Guided practice

Write 'True' or 'False' for each of these statements.

a Three-sixths is equivalent to one-half. <u>True</u>

b $\frac{3}{10} = \frac{1}{5}$ <u>False</u>

Number

Lesson 4: **Comparing and ordering fractions**

* Use knowledge of equivalence to compare and order proper fractions using the symbols =, > and <

Let's learn

We can compare and order **unit fractions** by looking at the denominators.
So $\frac{1}{5} > \frac{1}{7}$ and $\frac{1}{10} < \frac{1}{4}$.
These fractions have been ordered from smallest to greatest: $\frac{1}{8}$ $\frac{1}{5}$ $\frac{1}{3}$

But to compare and order **non-unit fractions**, we must consider the numerators, too. A fraction wall can help with this.

1							
$\frac{1}{2}$				$\frac{1}{2}$			
$\frac{1}{4}$		$\frac{1}{4}$		$\frac{1}{4}$		$\frac{1}{4}$	
$\frac{1}{8}$	$\frac{1}{8}$	$\frac{1}{8}$	$\frac{1}{8}$	$\frac{1}{8}$	$\frac{1}{8}$	$\frac{1}{8}$	$\frac{1}{8}$

To compare $\frac{3}{4}$ and $\frac{7}{8}$, look at their widths.
$\frac{7}{8} > \frac{3}{4}$ or $\frac{3}{4} < \frac{7}{8}$.

Change the fractions into fractions with the same denominator – equivalent fractions.

To order $\frac{3}{4}$ $\frac{7}{8}$ $\frac{1}{2}$ change them into eighths.
$\frac{3}{4} = \frac{6}{8}$ $\frac{1}{2} = \frac{4}{8}$
Now arrange them in order.
$\frac{4}{8}$ $\frac{6}{8}$ $\frac{7}{8}$ or $\frac{1}{2}$ $\frac{3}{4}$ $\frac{7}{8}$

Kim put these fractions in descending order: $\frac{7}{10}$ $\frac{1}{2}$ $\frac{4}{5}$
Is she correct?
Talk to your partner about what you think and how you know.

Number

Lesson 1: **Fractions as division**

- Understand that a fraction can be shown as a division

Key words
- fraction
- whole
- part
- equal
- equivalent
- divide
- share
- denominator
- numerator

Let's learn

We can show a fraction as a division.

$1 \div 2$ is equivalent to $\frac{1}{2}$.

$\frac{1}{2}$	$\frac{1}{2}$

$1 \div 3$ is equivalent to $\frac{1}{3}$.

$\frac{1}{3}$	$\frac{1}{3}$	$\frac{1}{3}$

$1 \div 5$ is equivalent to $\frac{1}{5}$.

$\frac{1}{5}$	$\frac{1}{5}$	$\frac{1}{5}$	$\frac{1}{5}$	$\frac{1}{5}$

$1 \div 8$ is equivalent to $\frac{1}{8}$.

$\frac{1}{8}$	$\frac{1}{8}$	$\frac{1}{8}$	$\frac{1}{8}$	$\frac{1}{8}$	$\frac{1}{8}$	$\frac{1}{8}$	$\frac{1}{8}$

Molly, Priya, Adnan and Ben share these two biscuits equally.
What fraction of each biscuit will they eat?
In total, what fraction of one biscuit will they eat?
Talk about this with your partner.
Explain how you know.

Guided practice

Write the division and the fraction that matches the bar.

$\frac{1}{4}$	$\frac{1}{4}$	$\frac{1}{4}$	$\frac{1}{4}$

$$1 \div 4 = \frac{1}{4}$$

Lesson 2: **Fractions of quantities**

Number

- Understand that fractions can act as operators

Let's learn

This shape has been divided into 9 equal parts.

$1 \div 9 = \frac{1}{9}$

$\frac{1}{9}$	$\frac{1}{9}$	$\frac{1}{9}$	$\frac{1}{9}$	$\frac{1}{9}$	$\frac{1}{9}$	$\frac{1}{9}$	$\frac{1}{9}$	$\frac{1}{9}$

This is $\frac{1}{9}$ of the whole. $\boxed{\frac{1}{9}}$

To find $\frac{1}{9}$ of 36, we can share 36 into the 9 parts.

$\frac{1}{9}$ of 36 = 4

We can use multiplication and division facts.

Knowing $4 \times 9 = 36$ tells us that $36 \div 9 = 4$. So $\frac{1}{9}$ of 36 = 4.

It also means that $9 \times 4 = 36$ and $36 \div 4 = 9$. So $\frac{1}{4}$ of 36 = 9.

28			
7	7	7	7

If you know that $\frac{1}{4}$ of 28 is 7, what is the value of $\frac{3}{4}$ of 28?

To find this, you need three lots of 7.

Think of some other whole numbers that you can find $\frac{1}{4}$ of.

Write the value of each quarter. Then work out the value of $\frac{3}{4}$ of each.

Guided practice

Find the value of each of these. Then write what else you know.

a $\frac{1}{3}$ of 18 $\boxed{6}$ $\boxed{\frac{1}{6} \text{ of } 18 = 3}$ **b** $\frac{1}{5}$ of 40 $\boxed{8}$ $\boxed{\frac{1}{8} \text{ of } 40 = 5}$

67

Number

Lesson 3: **Adding fractions**

- Add fractions with the same denominator

Let's learn

Adding fractions with the same denominator is like adding whole numbers but we only add the numerators.

$$\frac{3}{7} \qquad \frac{2}{7}$$

$\frac{1}{7}$	$\frac{1}{7}$	$\frac{1}{7}$	$\frac{1}{7}$	$\frac{1}{7}$	$\frac{1}{7}$	$\frac{1}{7}$

$$\frac{5}{7}$$

$3 + 2 = 5$, so:

$$\frac{3}{7} + \frac{2}{7} = \frac{5}{7}$$

The denominator stays the same because the number of parts does not change.

Commutativity works with fractions just as with whole numbers. That means $\frac{3}{8} + \frac{2}{8} = \frac{5}{8}$ then, $\frac{2}{8} + \frac{3}{8} = \frac{5}{8}$.

$$\frac{3}{8} \qquad \frac{2}{8}$$

$\frac{1}{8}$	$\frac{1}{8}$	$\frac{1}{8}$	$\frac{1}{8}$	$\frac{1}{8}$	$\frac{1}{8}$	$\frac{1}{8}$	$\frac{1}{8}$

$$\frac{5}{8}$$

Write five fractions that have a denominator of 9.

Choose pairs of them to add together.

Now choose three to add together.

Now add all of them together.

Talk to your partner about how you have done this.

Guided practice

Add these fractions.

a $\frac{1}{10} + \frac{3}{10} = \dfrac{4}{10}$　　　**b** $\frac{4}{7} + \frac{1}{7} = \dfrac{5}{7}$　　　**c** $\frac{3}{12} + \frac{5}{12} = \dfrac{8}{12}$

Lesson 4: **Subtracting fractions**

- Subtract fractions with the same denominator

Let's learn

Subtracting fractions with the same denominator is like subtracting whole numbers but we only subtract the numerators.

$9 - 2 = 7$, so: $\frac{9}{10} - \frac{2}{10} = \frac{7}{10}$

The denominator stays the same because the number of parts does not change.

If we know one fact, we also know another one.

$9 - 7 = 2$, so: $\frac{9}{10} - \frac{7}{10} = \frac{2}{10}$

Because of the relationship between addition and subtraction, we can write a family of four facts from just one.

$\frac{9}{10} - \frac{7}{10} = \frac{2}{10}$, $\frac{9}{10} - \frac{2}{10} = \frac{7}{10}$, $\frac{7}{10} + \frac{2}{10} = \frac{9}{10}$ and $\frac{2}{10} + \frac{7}{10} = \frac{9}{10}$

Write five fractions that have a denominator of 8.

Choose pairs of them to subtract.

For each one, write the other subtraction and the two additions that you know.

Talk to your partner about how you have done this.

Guided practice
Subtract these fractions.

a $\frac{7}{10} - \frac{3}{10} = \frac{4}{10}$

b $\frac{6}{7} - \frac{1}{7} = \frac{5}{7}$

c $\frac{10}{12} - \frac{3}{12} = \frac{7}{12}$

Number

Lesson 1: **What is a percentage? (1)**

- Understand that a percentage is part of a whole

Let's learn

We see **percentages** in everyday life.
We can see them in the shops.

20% tells us the amount of money we will save on an item we may want to buy.

We also see percentages on tablets and computers.

These percentages show us how much battery charge is left.

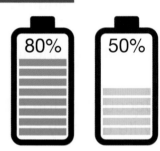

A percentage is an amount out of 100.

100% is the whole amount. It is $\frac{100}{100}$. 100% is also equal to 1.

20% is $\frac{20}{100}$ 50% is $\frac{50}{100}$ 80% is $\frac{80}{100}$

Look at the two batteries.

Talk to your partner about how much battery has been used and how much is left.

Which battery needs charging up first?

Guided practice

Shade these shapes to show 100%.

a

b

Number

Lesson 2: **What is a percentage? (2)**

- Understand that a percentage is part of a whole

Key words
- per cent
- percentage
- fraction
- whole
- part
- equivalent

Let's learn

We see percentages on clothes labels. These labels tell us the percentage of different fabrics in the clothes.

This grid is made up from 100 squares.
The whole grid represents 100%.

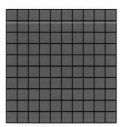

50 of the 100 squares on this grid is red.
Half this grid is red.
50% of the grid is red.

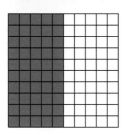

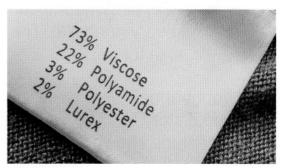

Look at the grid that shows 50%.

What percentage do you need to add to this to make 100%?

Talk to your partner about how you know.

If there was only 10% shaded, what percentage would you need to add to make 100%?

Make up some addition statements to show two percentages that make 100%.

Guided practice

Write the missing percentage to make 100%.

a $100\% = 20\% + \boxed{80\%}$ **b** $100\% = 5\% + \boxed{95\%}$

71

Number

Lesson 3: **Expressing hundredths as percentages**

• Understand that a percentage is part of a whole

Let's learn

Percentages are like fractions.

They tell us the number of parts in every hundred.

There are 100 people in the picture.
One person is wearing a hat.

We can say that $\frac{1}{100}$ of the people are wearing a hat.

We can also say that 1% of the people are wearing a hat.

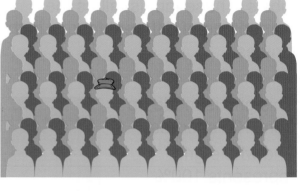

 Per cent means 'for every hundred'. The sign for per cent is %.

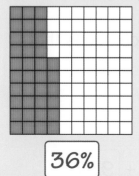 With your partner, make up a story for the percentage shown.

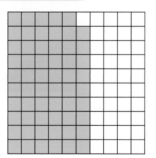

If you have 3 people out of every 100, you write 3%; for 17 people out of every 100, you write 17%.

Guided practice

The shaded part of each grid represents a percentage.
Write the percentage.

a

36%

b

62%

72

Lesson 4: **Expressing fractions as percentages**

Key words
- per cent
- percentage
- fraction
- whole
- part
- hundredth
- equivalent

Number

- Understand that fractions can be represented as percentages

Let's learn

A percentage can be thought of as another name for hundredths. A fraction expressed as a hundredth can simply be renamed as a percentage.

For example: $\frac{1}{100} = 1\%$ $\frac{27}{100} = 27\%$

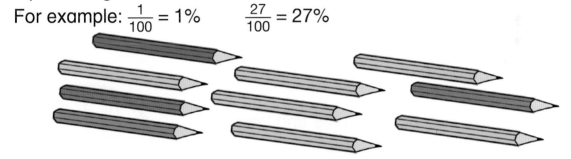

$\frac{6}{10}$ of these pencils are yellow. This is the same as $\frac{60}{100}$.
So, we can say that 60% of these pencils are yellow.

$10\% = \frac{10}{100}$, which is the same as $\frac{1}{10}$.
$50\% = \frac{50}{100}$, which is the same as $\frac{1}{2}$.

$25\% = \frac{25}{100}$, which is the same as $\frac{1}{4}$.
$75\% = \frac{75}{100}$, which is the same as $\frac{3}{4}$.

Make up a story like the one about the pencils.

Use a different number of tenths.

Convert the tenths to hundredths and the equivalent percentage.

Guided practice

Shade the fraction of the grid shown and write the percentage.

$\frac{25}{100} = \boxed{25\%}$

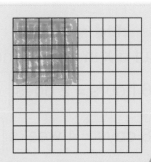

Lesson 1: **Units of time**

- Understand the relationship between units of time, and convert between them

Key words
- **seconds**
- **minutes**
- **hours**
- **days**
- **weeks**
- **months**
- **years**
- **units of time**

Let's learn

60 seconds (s) = 1 minute (min)
60 minutes = 1 hour (h)
24 hours = 1 day
7 days = 1 week
52 weeks = 1 year
12 months = 1 year
Some months have
30 days, others have 31 days.
February has 28 or 29 days.

We can use these facts, with mental calculation strategies, to work out lots of other facts.

This clock shows seconds, minutes and hours.

This clock reads
9 minutes and
1 second past 10.

Geometry and Measure

 What is the same about the calendar pages for February 2019 and February 2020?
What is different?

Talk to your partner about what you notice.

Write as many similarities and differences as you can see.

| **February 2019** | | | | | | |
M	T	W	T	F	S	S
				1	2	3
4	5	6	7	8	9	10
11	12	13	14	15	16	17
18	19	20	21	22	23	24
25	26	27	28			

| **February 2020** | | | | | | |
M	T	W	T	F	S	S
					1	2
3	4	5	6	7	8	9
10	11	12	13	14	15	16
17	18	19	20	21	22	23
24	25	26	27	28	29	

Guided practice

Put a ✗ in the box next to the statements that are incorrect.

46 hours = 2 days ✗

36 months = 3 years ☐

28 days = 4 weeks ☐

160 seconds = 3 minutes ✗

Lesson 2: **Linking analogue and 12-hour digital times**

- Read and record time accurately on analogue and 12-hour digital clocks

Let's learn

Noon is 12 o'clock, in the middle of the day. It is also called midday.

We can write a.m. after morning times. This stands for **ante meridiem**, which means before noon.

We can write p.m. after afternoon and evening times. This stands for **post meridiem**, which means after noon.

14 minutes past 8 in the evening

42 minutes past 5 in the morning is the same as 18 minutes to 6 in the morning.

Draw these times on three of your analogue clocks.

   a 37 minutes past 6 in the morning

   b 37 minutes past 11 in the morning

   c 37 minutes past 2 in the afternoon

Write the digital time underneath each clock.

Don't forget to write a.m. or p.m.

Work with your partner to find the interval, in hours, between times **a** and **b**, **b** and **c**, **a** and **c**.

Guided practice

Write an equivalent analogue time beside each digital time.

9:42 | 42 minutes past 9

3:38 | 38 minutes past 3

6:21 | 21 minutes past 6

Geometry and Measure

75

Lesson 3: **Linking 12- and 24-hour time**

Key words
- minutes
- hours
- analogue clock
- hands
- digital clock
- a.m.
- p.m.
- 12-hour time
- 24-hour time

- Read and record time accurately on analogue and 12- and 24-hour digital clocks

Let's learn

This is 24-hour time. It does not use a.m. or p.m. This analogue clock shows the hours from 1 to 12, then from 13 to 23. At midnight the numbers start again at 00.

7:00 a.m. is 07:00. So 7:18 a.m. will be 07:18.

9:00 a.m. is 09:00. So 9:36 a.m. will be 09:36.

1:00 p.m. is 13:00. So 1:25 p.m. will be 13:25.

9:00 p.m. is 21:00. So 9:54 p.m. will be 21:54.

12:00 (midnight) is 00:00. So 12:43 a.m. will be 00:43.

12:00 (noon) is 12:00. So 12:12 p.m. will be 12:12.

Look at the 24-hour analogue clock.

Talk to your partner about how you would read the time from it.

Write three 12-hour times then swap with your partner and write their times as 24-hour times.

Repeat, but start with 24-hour times.

Guided practice

Write 'True' or 'False' for each statement.

a 18:36 is the same as 6:36 a.m. False

b 10 minutes to 7 in the morning is the same as 06:50. True

Lesson 4: **Timetables**

- Interpret and use timetables in 12- and 24-hour times

Key words
- minutes
- hours
- a.m.
- p.m.
- 12-hour time
- 24-hour time
- timetable
- interval

Geometry and Measure

Let's learn

We can use time number lines to calculate time intervals.

This time number line is marked in 5-minute intervals.

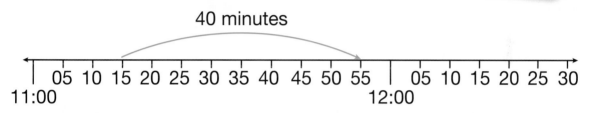

If we can work out time intervals, we can read and use timetables.

Bus stop	Arrival time		
Market Street	10:15	10:35	10:55
River Street	10:25	10:45	11:05
Easton	10:35	10:55	11:15
School Road	10:45	11:05	11:25
Westerby	10:55	11:15	11:35

Look at the bus timetable above.

It takes 10 minutes to travel from Market Street to River Street.

Work with your partner to find six more facts from the timetable.

Guided practice

Calculate the time difference. Draw a time line to help you.

7:15 a.m. to 7:56 a.m.   41 minutes

15 minutes   26 minutes

7:15   7:30   7:56

Lesson 1: **Combining polygons**

- Investigate shapes made by combining two or more shapes

Let's learn

How many shapes can you see in the window?

The shapes all have straight sides so they are **polygons**.

The windows combine to make an octagon.

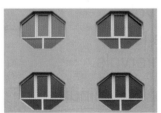

> We can put shapes together to make other shapes.

These two triangles make a rectangle.

These two pentagons make an octagon.

When we draw some shapes next to each other so that there are no gaps, we make a **tessellating pattern**.

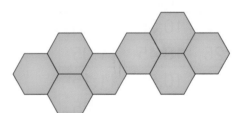

Here is a group of green shapes and a group of blue shapes.

5 How could we classify them, apart from by colour?

Talk to your partner about what you think.

Guided practice

Here are three hexagons.

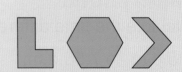

a Write two ways they are alike.

> They all have 6 sides and 6 vertices.

b Write two ways they are different.

> The lengths of their sides are different and the sizes of their vertices are different.

Geometry and Measure

78

Lesson 2: **Combining 4-sided shapes**

• Combine two or more 4-sided shapes

Let's learn

A square is the only regular 4-sided shape.

Rectangles and squares have four right angles.

Squares are rectangles with four equal sides.

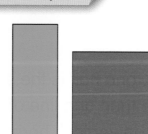

Here are some more 4-sided shapes.

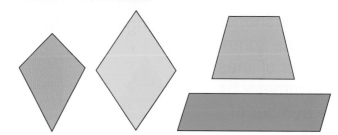

Some of these shapes have pairs of parallel sides.

We can put 4-sided shapes together to make other shapes.

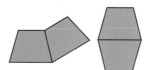

These are hexagons.   This is an octagon.

• You and your partner each think of a different 4-sided shape.

Draw your shapes on squared paper and cut them out. Try to make your shapes the same size.

Then put the two shapes together to make new shapes.

Sketch the shapes you can make.

Geometry and Measure

Guided practice

Sketch a regular 4-sided shape that has parallel sides.
Write its properties.

It has 4 sides, the opposite two pairs are parallel and it has 4 right angles.

Lesson 3: **Symmetry of 2D shapes**

- Identify horizontal, vertical and diagonal lines of symmetry

Key words
- **symmetry**
- **line of symmetry**
- **vertical**
- **horizontal**
- **diagonal**

Let's learn

In a symmetrical shape, one half is the same as the other half.

A shape can have more than one line of symmetry.

A line of symmetry can be vertical, horizontal or diagonal.

Regular shapes have the same number of lines of symmetry as the number of sides.

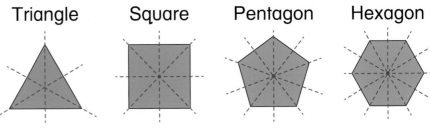

Triangle   Square   Pentagon   Hexagon

three lines of symmetry   four lines of symmetry   five lines of symmetry   six lines of symmetry

Squares have 4 sides and 4 lines of symmetry. On this square, one line is horizontal, one is vertical and two are diagonal.

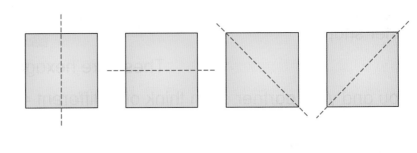

👥 Each shape is made from five squares.

With your partner, identify the lines of symmetry each shape has.

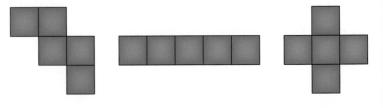

Copy them onto squared paper. Draw on the lines of symmetry.

Guided practice

Inside each shape write the number of lines of symmetry it has.

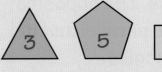

3   5   2

Lesson 4: **Symmetry in real life**

- Identify horizontal, vertical and diagonal lines of symmetry

Key words
- symmetry
- line of symmetry
- mirror line
- horizontal
- vertical
- diagonal
- reflect

Let's learn

This photograph shows the reflection of a tree in water. The image we see is symmetrical. The line of symmetry is horizontal.

A reflection is a shape or an image seen as it would be in a mirror. A mirror helps you to make shapes or images symmetrical.

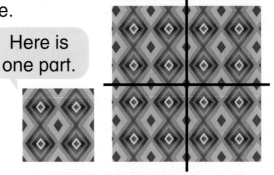

The mirror lines on this pattern divide the pattern into four parts that are the same.

We could take one part and copy it three times and then put the parts together to make the pattern.

> Here is one part.

Can you see how to fit four of these parts together to make the pattern?

Talk to your partner about the mirror lines in this pattern and describe the reflections they make.

Make a pattern like this on squared paper.

Give your pattern four lines of symmetry.

Draw on the mirror lines.

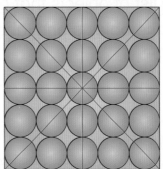

Geometry and Measure

Guided practice

Draw the lines of symmetry on these flags.

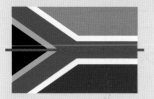

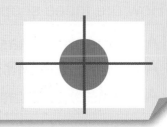

Lesson 1: **Shapes with curved surfaces**

Geometry and Measure

• Identify 2D faces of 3D shapes and describe their properties

Key words
• sphere
• hemisphere
• cylinder
• cone
• curved surface
• circular face
• circular edge
• vertex
• apex

Let's learn

We can sort 3D shapes into two groups:

• shapes that have flat faces, straight edges and vertices

• shapes that have at least one curved surface.

All flat faces	Not all flat faces

A **sphere** has one curved surface.

A **cylinder** has one curved surface, two circular faces and two circular edges.

A **cone** has one curved surface, one circular face, one circular edge and the point is called an apex.

This marble is a sphere.

This drum is a cylinder.

This party hat is a cone.

Make a list of objects that are spheres, cylinders and cones. Sketch the objects and label them with the name for the shape.

Guided practice

How are these shapes alike?

They both have curved surfaces.

How are these shapes different?

A hemisphere has a circular edge and a circular face. A sphere doesn't. A sphere can roll in any direction. A hemisphere can only roll from side to side.

Lesson 2: **Prisms**

- Identify different prisms and describe their properties

Key words
- triangular prism
- cube
- cuboid
- pentagonal prism
- hexagonal prism
- face
- edge
- vertex
- vertices

Let's learn

Prisms are 3D shapes with two identical flat faces joined by rectangles. The flat faces must be polygons (2D shapes with straight sides).

Prisms	Not prisms

A **triangular prism** has 2 triangular faces and 3 rectangular faces. It has 9 edges and 6 vertices.

A **cube** has 6 faces that are all squares. It has 12 edges and 8 vertices.

A **cuboid** has 6 faces that are all rectangles. The two end faces can be squares. It has 12 edges and 8 vertices.

There are lots of different prisms.

What would a pentagonal prism look like?
Work out the number of faces, edges and vertices.
What about a hexagonal prism?

Guided practice

Tommy thinks that the first shape is a triangular prism but not the second shape. Do you agree? No
Explain.

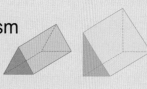

> A prism has two identical ends joined by rectangles. The second is a triangular prism because it has two identical triangles joined by rectangles.

Geometry and Measure

Lesson 3: **Pyramids**

- Identify different pyramids and describe their properties

Let's learn

Pyramids are 3D shapes with a base and triangular faces that are joined to the base. The base must be a polygon (2D shape with straight sides).

Pyramids	Not pyramids

Key words
- **triangular-based pyramid**
- **tetrahedron**
- **square-based pyramid**
- **pentagonal-based pyramid**
- **hexagonal-based pyramid**
- **face**
- **edge**
- **vertex**

A **triangular pyramid** has 4 triangular faces.
It is also called a **tetrahedron**.
It has 6 edges and 4 vertices.

A **square-based pyramid** has one square face and 4 triangular faces.
It has 8 edges and 5 vertices.

Any polygon can be the base of a pyramid.

With your partner, write five different examples of pyramids and describe their properties.

Guided practice

Ekon thinks the first shape is a triangular-based pyramid but not the second shape because he can only see two triangles.

Do you agree? | No |
Explain.

> All triangular-based pyramids have four triangular faces. The second has two faces that can't be seen. One is the base and the other is at the back.

Lesson 4: **Nets**

• Match nets to their corresponding 3D shapes

Let's learn

A **net** is a 2D shape that can be folded to make a 3D shape.

Each 3D shape has a different net.

If we know the properties of a 3D shape we can work out how to draw its net.

A prism has two ends that are polygons, which are joined by rectangles.

Nets of a cube

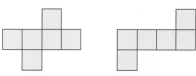

Net of a squared-based pyramid

Nets of a cuboid

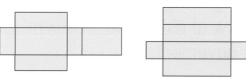

A pyramid has a base that is a polygon. The other faces are triangles.

Talk to your partner about this net.

What 3D shape will it make?

Sketch the net on paper, cut it out and fold it to see if you are correct.

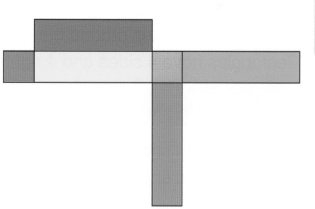

Geometry and Measure

Guided practice

Draw lines to match each net to its shape.

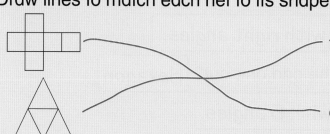

triangular-based pyramid

cube

Lesson 1: **Angles all around us**

Key words
• angle
• degrees
• right angle

• Recognise angles

Let's learn

In a 2D shape, two sides meet at a vertex to form an **angle**.

The angle is the amount of turn between the sides.

We measure angles in **degrees**.

Angles can be different sizes.

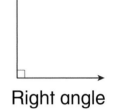

Greater than a right angle

Right angle

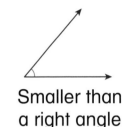

Smaller than a right angle

A right angle is the same size as a quarter turn.

Look at the angles in this picture of part of a pavement.

Can you see any right angles?

Can you see any angles that are smaller than a right angle?

What about any that are greater?

Look for right angles around the classroom.

Rectangles have 4 sides and 4 right angles.

Draw a 4-sided shape that has 2 right angles.
Draw a 4-sided shape that has no right angles.

Geometry and Measure

Guided practice

Which shape is the odd one out? | Square |

Why? | It is the only shape with right angles. |

Which other shape could be the odd one out? | Hexagon |

Why? | It is the only shape with 6 angles. |

Lesson 2: **Right angles**

- Identify and combine right angles

Let's learn

Here are some right angles.

Key words
- **angle**
- **degrees**
- **right angle**
- **quarter turn**
- **half turn**
- **three-quarter turn**
- **full turn**

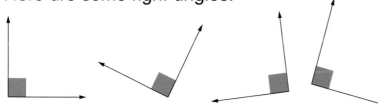

It doesn't matter how they are positioned; they are always right angles if the angle is a quarter turn.

A **right angle** measures 90°.	A **straight line** or half a turn is made from two right angles and measures 180°.	A **three-quarter turn** is made from three right angles and measures 270°.	A **whole turn** is made from four right angles and measures 360°.

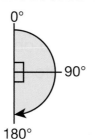

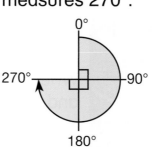

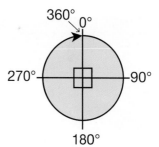

Draw two squares, arranged to show a straight line.
Label with the size of the straight line.

Draw three squares, arranged to show a three-quarter turn.
Label with the size of the turn.

Draw four squares, arranged to show a whole-turn.
Label with the size of the turn.

Geometry and Measure

Guided practice

How many degrees are there in half a turn? 180°

How many right angles is this equivalent to? 2 right angles

Lesson 3: **Acute and obtuse angles**

• Identify acute and obtuse angles

Let's learn

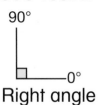

Right angle

Less than a right angle (acute)

Greater than a right angle (obtuse)

A **right angle** is 90°.

Angles that are smaller than a right angle are **acute angles**.

They measure greater than 0° but less than 90°.

Angles that are greater than a right angle but smaller than a straight line are **obtuse angles**.

An obtuse angle is greater than 90° and less than 180°.

Look at this stick insect's legs.

Can you see acute and obtuse angles?

Where else in the photo can you see acute and obtuse angles?

This irregular hexagon has three right angles, one acute angle, one obtuse angle and one angle that is three-quarters of a turn.

Are the five angles inside a regular pentagon acute angles, right angles or obtuse angles?

Sketch a regular pentagon to prove what you think.

Sketch an irregular pentagon that has at least one acute angle, one right angle and one obtuse angle.

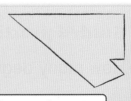

Guided practice

Draw a hexagon with two right angles.

What can you say about the other angles?

Two of the angles are acute, one is obtuse and the other is greater than an obtuse angle.

Geometry and Measure

Lesson 4: **Comparing and ordering angles**

> **Key words**
> * angle
> * degrees
> * right angle
> * acute angle
> * obtuse angle

• Estimate, compare, order and classify angles

Let's learn

Acute angles are greater than 0° but less than 90°.

Right angles measure 90°.

Obtuse angles are greater than 90° and less than 180°.

We can use these facts and the symbols < and > to compare and order angles.

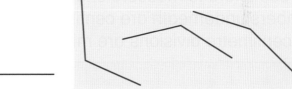

If you are not sure, use a right angle to help you compare.

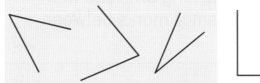

Smaller than a right angle   Greater than a right angle

Draw an angle and place the less than or greater than symbol to the right of your angle.

Swap with your partner and draw an angle to make the statement true.

Repeat several times.

Guided practice

Order these angles, from smallest to greatest.

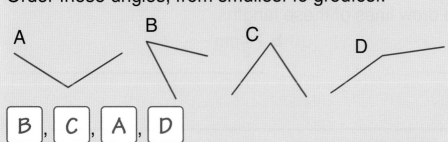

| B | , | C | , | A | , | D |

Geometry and Measure

Lesson 1: **Rulers**

• Read and interpret scales on a ruler

Let's learn

We can measure lengths in millimetres, centimetres, metres and kilometres.

This boy is using a metre stick to measure the width of the window.

It is $\frac{1}{2}$ a metre wide. $\frac{1}{2}$ m is the same as 50 cm.

The window is $\frac{1}{2}$ metre.

We use rulers to measure short lengths. The long divisions with numbers underneath are centimetres. The small marks between the centimetre divisions are millimetres.

Use a ruler to draw five lines of different lengths.

Swap your lines with your partner. Estimate how long the lines are.

Now measure with a ruler.

How close were your estimates?

Guided practice

Use your ruler to draw lines of these lengths.

a $4\frac{1}{2}$ cm

b $7\frac{1}{2}$ cm

_____

_____

c $14\frac{1}{2}$ cm

_____

Lesson 2: **Finding mass**

• Use scales to find mass

Let's learn

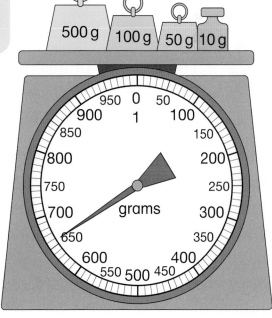

We can use bathroom scales, kitchen scales and balance scales to find the mass of something.

These balance scales show the mass of a book. It has a mass of $\frac{1}{2}$ kg, which is the same as 500 g.

• Talk to your partner about the mass on the kitchen scales.

How much heavier than 600 g is the mass?

How much lighter is it than 700 g?

Geometry and Measure

Guided practice
How many grams?

a 1 kg `1000 g`

b $1\frac{1}{2}$ kg `1500 g`

c $\frac{1}{4}$ kg `250 g`

Lesson 3: **Measuring cylinders**

* Read and interpret scales on measuring cylinders

Let's learn

These measuring instruments all show 100 m*l*.

100 m*l* of liquid looks different in different containers.

100 m*l* is $\frac{1}{10}$ of a litre.

100 m*l*

100 m*l*

100 m*l*

100 m*l*

The capacity of this measuring cylinder is 1 *l* or 1000 m*l*.

It is less than half full.

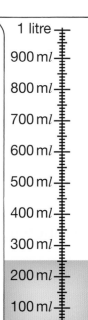

👥 Look again at the measuring cylinder. How many millilitres of liquid are in it?

How can we say that in litres?

With your partner, sketch two measuring cylinders on paper.

Show one of them with $\frac{1}{2}$ *l* of liquid in it and the other with $\frac{3}{4}$ *l*.

Guided practice

How many millilitres?

a $1\frac{1}{2}$ *l* 1500 m*l*

b $2\frac{1}{4}$ *l* 2250 m*l*

c $3\frac{1}{2}$ *l* 3500 m*l*

Geometry and Measure

Lesson 4: **Revising measurement**

- Read and interpret scales for length, mass and capacity

Let's learn

This pencil is $25\frac{1}{2}$ cm long.
We could say that its length is 255 mm.

The mass of this watermelon is $4\frac{1}{4}$ kg.
We could say that its mass is 4 kg and 250 g or 4250 g.

The capacity of this petrol can is 4300 m*l*.
We could also say that its capacity is 4 *l* 300 ml or 4 and $\frac{3}{10}$ of a litre.

4300 m*l*

With your partner, think of something that is measured in kilometres. Write it down.

Now think of something that is measured in kilograms and something that is measured in grams. Write them down.

Do the same for litres and millilitres and centimetres and millimetres.

Guided practice

Write each mass in grams.

a $\frac{1}{2}$ of a kilogram [500] g

b $\frac{1}{4}$ of a kilogram [250] g

c $\frac{1}{10}$ of a kilogram [100] g

d $\frac{3}{4}$ of a kilogram [750] g

Geometry and Measure

93

Lesson 1: **Perimeter and area of 2D shapes**

Key words
• perimeter
• area

• Find the perimeter and area of 2D shapes

Let's learn

6 cm

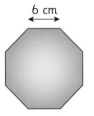

We can find the perimeter of a shape by measuring the length of each side and finding the total.

The calculation for the perimeter of this regular octagon is:

6 cm + 6 cm + 6 cm + 6 cm + 6 cm + 6 cm + 6 cm + 6 cm

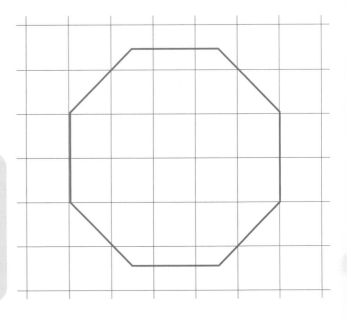

 We can find the area of a shape by counting whole and part squares. This octagon has an area of about 20 squares.

With your partner, find the perimeter of the octagon above.

2 Can you find a quicker way than adding the same number over and over again?

Guided practice

Find the perimeter of a regular triangle with sides of length:

a 5 m   15 m   **b** 8 cm   24 cm   **c** 12 mm   36 mm

Geometry and Measure

Lesson 2: **Perimeter and area of rectangles and squares**

Key words
- **perimeter**
- **area**
- **width**
- **length**
- **square centimetres (cm²)**

- Find the perimeter and area of rectangles and squares

Let's learn

To find the **perimeter** of this rectangle, we can add 2 cm and 11 cm and double.

In this example, 11 cm + 2 cm = 13 cm and 13 cm × 2 = 26 cm.

To find the perimeter of a square we can multiply the length of one side by 4.

To find the **area** of a rectangle we multiply the length by the width.

The area of the red rectangle is 11 cm × 2 cm = 22 cm².

To find the area of a square we multiply the side length by itself.

The area of the yellow square is 8 cm × 8 cm = 64 cm².

Look at the green rectangle.

Work out the perimeter.

Next work out the area.

Explain to your partner how you did this.

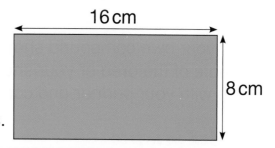

Geometry and Measure

Guided practice

Find the area and perimeter of a rectangle with these dimensions:

a 7 cm by 2 cm   Area = $14 \, cm^2$ Perimeter = $18 \, cm$

b 10 cm by 8 cm   Area = $80 \, cm^2$ Perimeter = $36 \, cm$

Lesson 3: **Finding the area of 2D shapes (1)**

• Find the area of compound shapes

Geometry and Measure

Let's learn

Compound shapes are made from simple shapes that have been joined together. In these examples, the two shapes that have been joined are rectangles.

To find the area of compound shapes, we find the areas of the two rectangles and add them together.

Here the two rectangles are 2 cm × 2 cm and 3 cm × 10 cm. The two areas are 4 cm² and 30 cm² making a total area of 34 cm².

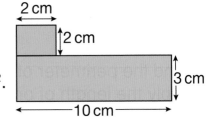

In this shape the two rectangles are 2 cm × 5 cm and 8 cm × 3 cm. The two areas are 10 cm² and 24 cm² which also give a total area of 34 cm².

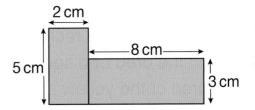

👥 Draw your own compound shape on squared paper. Make a sensible estimate of the area of your shape. Then find the actual area. Swap with your partner and compare your results.

Guided practice

What is the area of this shape? Show how you worked it out.

6 × 2 = 12
2 × 2 = 4
12 + 4 = 16
The area is 16 cm².

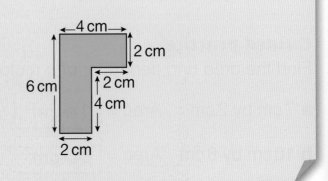

Lesson 4: **Finding the area of 2D shapes (2)**

Key words
- area
- square centimetres
- compound shape

- Find the area of irregular shapes

Geometry and Measure

Let's learn

Compound shapes are made by joining simple shapes.

These compound shapes have been made by joining three rectangles.

Can you see the three rectangles?

We can divide the shapes in different ways.

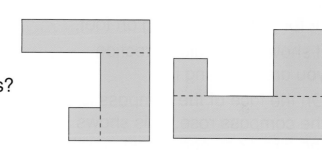

Some of the side lengths are missing from this compound shape. We use what we know to find the missing lengths.

Draw this shape on paper. Divide it into three rectangles. Find the area of each rectangle. Add the three areas together to find the total area.

12 m

8 m

9 m

4 m   4 m

Guided practice

Find the area of this compound shape

Area of 1st rectangle:

| $2\,cm \times 2\,cm = 4\,cm^2$ |

Area of 2nd rectangle:

| $2\,cm \times 8\,cm = 16\,cm^2$ |

Area of 3rd rectangle:

| $3\,cm \times 4\,cm = 12\,cm^2$ |

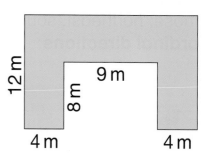

2 cm

1

6 cm

4 cm

2

5 cm

3 cm   3

4 cm

Total area: $4\,cm^2 + 16\,cm^2 + 12\,cm^2 = 32\,cm^2$

Geometry and Measure

Lesson 1: **Compass points**

- Use compass points to describe position, direction and movement

Key words
- compass
- cardinal
- north
- east
- south
- west
- direction
- ordinal
- northeast
- southeast
- southwest
- northwest

Let's learn

A **compass** is a very important instrument for a pilot of a ship or a plane or for people exploring on foot.

It shows you which direction you are travelling in.

On the face of the compass is the compass rose. This shows the directions north, east, south and west. These are called **cardinal directions**. It also shows the directions halfway between those: northeast, southeast, southwest and northwest. These are called **ordinal directions**.

north (N)
northwest (NW)   northeast (NE)
west (W)   east (E)
southwest (SW)   southeast (SE)
south (S)

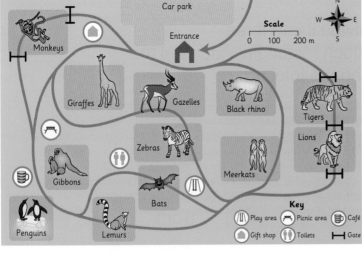

Look at this map of a wildlife park.

Imagine you are visiting the zebras.

From there, the black rhino are in a northeast direction.

The penguins are southwest of the zebras.

The giraffes are in a northwest direction.

To get to the meerkats you need to go southeast.

👥 Work with your partner.

Find the gibbons on the map.

Discuss in which direction you would travel from the gibbons to see other animals or features in the park.

Guided practice

Write two directions in which we could travel. Include their abbreviations.

north (N) and northeast (NE)

Lesson 2: **Position, direction and movement**

- Use the vocabulary of position, direction and movement

Let's learn

If we are going somewhere and do not know where it is, we might need to follow directions.

We can use compass directions to find our way.

We can use other words as well.

We can use words such as: left, right, forwards, backwards, up and down.

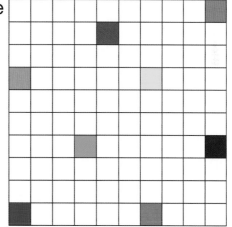

To travel from the red square to the green square we could move 6 spaces to the left. Then we could move down 6 squares.

How could you move from the red square to the pink square passing over each coloured square? How many ways can you find?

Guided practice

Look at the grid above. Write a set of instructions to travel from the blue square to the yellow square without passing over another coloured square.

> Move right 1 space then travel north 6 spaces, turn right and travel east 5 spaces.

Lesson 3: **Coordinates**

- Use coordinates to describe position

Let's learn

Look at the crosses on this grid.

To find the position of the red cross, we count along the horizontal axis from 0 to 1 and then up the vertical axis to 3.

The red cross is at (1, 3). The numbers (1, 3) are the **coordinates** of the red cross.

For the blue cross, we count along the horizontal axis from 0 to 3 and then up the vertical axis to 1.

The coordinates of the blue cross are (3, 1).

Look carefully at how the coordinates are written.

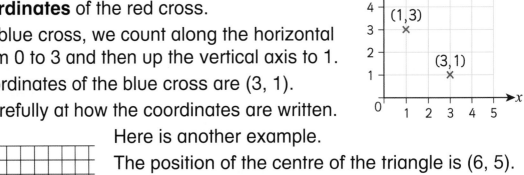

Here is another example.

The position of the centre of the triangle is (6, 5).

The horizontal axis is called the ***x*-axis**.

The vertical axis is called the ***y*-axis**.

The area of the graph between the two axes is called a **quadrant**.

Look at the coordinate grid below.

Together, work out where the six objects have been positioned.

Write the name of each object and its coordinates.

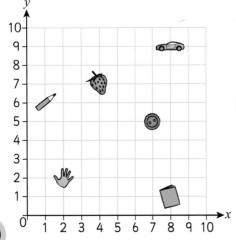

Guided practice

Draw five small shapes on this grid.
You decide where to put them.

List their coordinates.

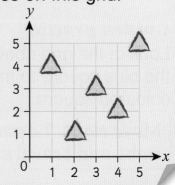

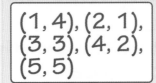
(1, 4), (2, 1),
(3, 3), (4, 2),
(5, 5)

Geometry and Measure

Lesson 4: **Identifying points**

- Identify points on a coordinate grid

Key words
- **coordinates**
- **x-coordinate**
- **y-coordinate**
- **horizontal axis**
- **x-axis**
- **vertical axis**
- **y-axis**

Let's learn

We identify points on a coordinate grid by reading the numbers on the axes.

We move along the horizontal x-axis first to find the x-coordinate.

Then we move up the vertical y-axis to find the y-coordinate.

The point of the pink arrow is at (6, 9).

It has an x-coordinate of 6 and a y-coordinate of 9.

The other end of the pink arrow has coordinates of (2, 9).

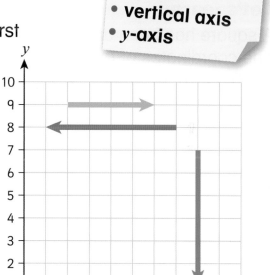

Look at the grid on the right.

Write the coordinates of both ends of the green arrow.

Then write the coordinates of both ends of the blue arrow.

What do you notice?

Guided practice

Write the coordinates for each star.

Blue (1, 4)
Yellow (3, 7)
Green (5, 5)
Orange (7, 8)
Red (8, 3)

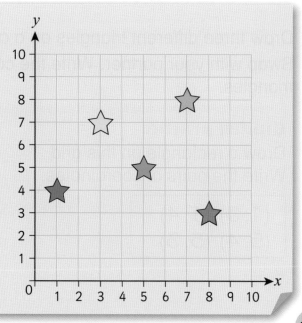

Lesson 1: **Reading and plotting coordinates**

Key words
• **coordinates**
• **horizontal axis**
• **x-axis**
• **vertical axis**
• **y-axis**
• **quadrant**

• Read and plot coordinates for vertices of shapes

Geometry and Measure

Let's learn

A square has been drawn on this grid. The coordinates of the square are (1, 2), (1, 5), (4, 5) and (4, 2).

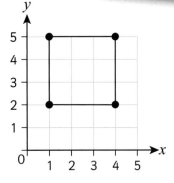

We can show any shape on a coordinate grid. Here are two examples. The coordinates for the triangle are (1, 2), (1, 5) and (4, 5).

The coordinates for the pentagon are (1, 2), (1, 5), (4, 5) (3, 3) and (2, 2).

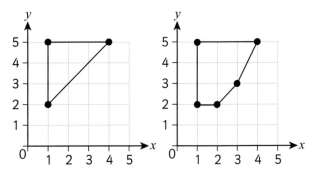

Draw three different triangles on a coordinate grid.

Swap with your partner. Write the coordinates of the vertices of their triangles.

Guided practice
Draw a rectangle on this grid.
Write the pairs of coordinates.

(1, 3) (1, 4)
(5, 4) (5, 3)

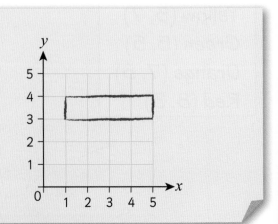

Lesson 2: **Shapes from coordinates**

- Read and plot coordinates for shapes

Let's learn

We can plot points and join them to draw shapes.

We can use coordinates to describe the positions of the points.

The coordinates of the triangle are (4, 5), (8, 5) and (4, 8).

The coordinates of the square are (1, 1), (1, 4), (4, 4) and (4, 1).

The coordinates of the rectangle are (6, 1), (6, 3), (10, 3) and (10, 1).

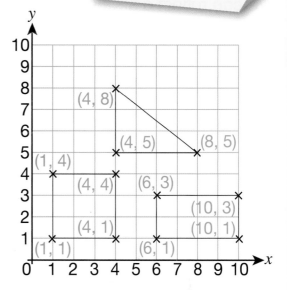

Draw a square on a coordinate grid. Give it to your partner. Your partner writes the coordinates for the vertices of your square.

Guided practice

Draw a square on the grid. Write all the coordinates of the square.

(1, 4), (4, 4), (4, 1) (1, 1)

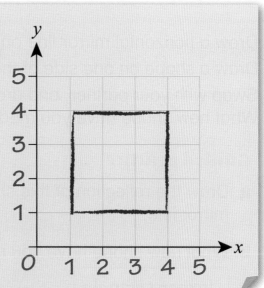

Geometry and Measure

Lesson 3: **Horizontal refection**

- Reflect 2D shapes in a horizontal mirror line

Let's learn

A **reflection** is an image of a shape as it would be seen in a mirror.

We can reflect a shape in one of its edges to create a new shape.

The red triangle has been reflected in a horizontal mirror line to make a 4-sided shape.

The orange square has been reflected in a horizontal mirror line to make a rectangle.

The green shape has been reflected in a horizontal mirror line to make a hexagon.

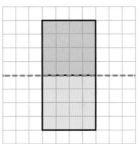
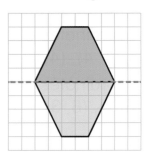

A mirror helps us to make shapes or images **symmetrical**. It also helps us to draw reflections.

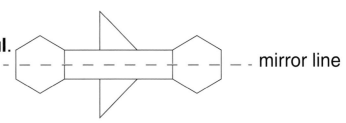

mirror line

👥 Draw a horizontal mirror line on a 10 × 10 grid.

Draw a shape on one side of the mirror line.

Swap with your partner, and draw the reflection of their shape. What new shape have you made?

Guided practice

a Draw the reflection of this shape in the mirror line.

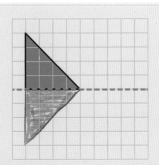

b What new shape have you made?

larger triangle

Lesson 4: **Vertical reflection**

• Reflect 2D shapes in a vertical mirror line

Key words
• reflection
• vertical
• line of symmetry
• mirror line

Let's learn

We can reflect shapes in vertical mirror lines.

It's just like reflection in a horizontal mirror line but the direction of reflection is different.

The purple triangle has been reflected in a vertical mirror line to make a 4-sided shape.

The yellow square has been reflected in a vertical mirror line to make a rectangle.

The orange shape has been reflected in a vertical mirror line to make a hexagon.

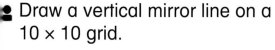

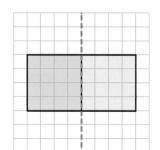

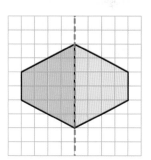

Draw a vertical mirror line on a 10 × 10 grid.

Draw a shape on one side of the mirror line.

Swap with your partner, and draw the reflection of their shape. Can you name the new shape you have made?

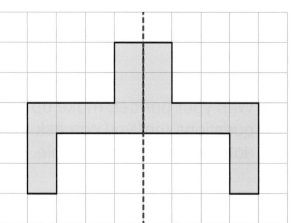

Guided practice

a Draw the reflection of this rectangle across the mirror line.

b What new shape have you made?

square

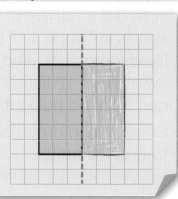

Geometry and Measure

Lesson 1: **Tally charts and frequency tables**

- Record, organise, represent and interpret data in tally charts and frequency tables

Let's learn

Data is information. Statistics is about collecting, sorting and representing information.

We use tallies to record results.

We can use tally charts and frequency tables to collect information.

To make a frequency table, we add a column to a tally chart. We count the tallies and write the total in the frequency column.

This frequency table shows **grouped data** about shoe sizes. 9 learners have shoes sized 1 and 2.

Where would your shoe size go on this table?

With your partner, make up five questions that you can ask from the data in the frequency table.

Shoe sizes	
Shoe sizes	**Frequency**
12–13	3
1–2	9
3–4	8
5–6	1

Guided practice

Ravi did a survey of favourite sports in Stage 4. 7 learners liked football, 9 liked swimming, 12 liked baseball and 14 liked basketball.

Show this information in a frequency table.

Sport	Tally	Frequency
football	卌 ‖	7
swimming	卌 ‖‖	9
baseball	卌 卌 ‖	12
basketball	卌 卌 ‖‖	14

Lesson 2: **Venn diagrams**

• Record, organise, represent and interpret data in Venn diagrams

Key words
• **statistics**
• **data**
• **Venn diagram**
• **interpret**
• **criterion**
• **criteria**

Let's learn

A **Venn diagram** shows sets of data.

The Venn diagram is divided into regions.

Each region contains data with a particular property.

The data in the regions where the circles overlap has two or more properties.

This Venn diagram compares apples, oranges and watermelons.

With your partner, write eight statements about the data on the Venn diagram.

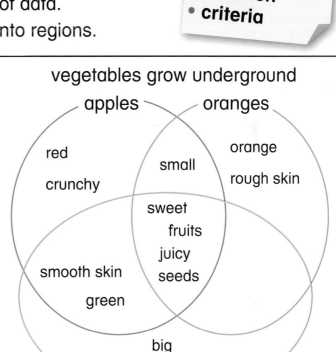

vegetables grow underground

apples — oranges

red
crunchy
small
orange
rough skin
sweet
fruits
juicy
smooth skin   seeds
green
big
heavy
watermelons

Guided practice

Write the numbers in the correct sections on the Venn diagram.

28 16 56 112 96 35

What does the overlap show?

The overlap shows that 56 and 112 are multiples of 7 and 8.

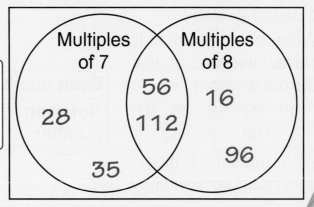

Multiples of 7   Multiples of 8

56
16
28
112
96
35

Lesson 3: **Carroll diagrams**

- Record, organise, represent and interpret data in Carroll diagrams

Key words
- **statistics**
- **data**
- **Carroll diagram**
- **interpret**
- **criterion**
- **criteria**

Let's learn

	Circular	Not circular
Green		
Not green		

Use a **Carroll diagram** to organise and group data according to whether it fits certain criteria.

Venn diagrams and Carroll diagrams show what is the same and what is different about sets of data we collect.

In a Carroll diagram, we sort by a criterion and **not** that criterion.

With your partner, write eight statements about the data on the Carroll diagram.

This Carroll diagram shows 'triangle' and 'not triangle', 'right angles' and 'not right angles'.

	Triangle	Not triangle
Right angles		
Not right angles		

Guided practice

Write the numbers in the correct sections on the Carroll diagram.

40 23 58 75 91
100 125 62

	Multiple of 5	Not multiple of 5
Even number	40 100	58 62
Not even number	75 125	23 91

What does the bottom right hand section show us?

That 23 and 91 are not multiples of 5 and not even numbers.

Statistics and Probability

108

Lesson 4: **Pictograms**

> • Record, organise, represent and interpret data in pictograms

Let's learn

We can use a **pictogram** to organise data.

The symbol represents the numerical information and equals the number in the key.

A half symbol represents half the quantity of the key.

When choosing a symbol to use for a pictogram, choose a symbol that can be divided easily.

With your partner, work out how many of each coloured sweet there are in the bag.

Then work out the total number of sweets there are in the bag. Together, write five statements about the data on the pictogram.

What's in the bag of sweets?

green	● ● ● ● (
orange	● ● ● ●
blue	● ● (
pink	● ● ●
yellow	● ● ● ● ● (
red	● ● ● ●
purple	● ● ● (
brown	● (

Key: ● = 2 sweets

Guided practice

Four learners were looking for insects in the school nature garden. They recorded the results in a frequency table.

Name	Frequency
Martha	40
Akiba	20
Ruth	25
Adnan	10

Martha	▲ ▲ ▲ ▲
Akiba	▲ ▲
Ruth	▲ ▲ ◢
Adnan	▲

Key ▲ = 10 insects

Show these results in a pictogram.

Statistics and Probability

Lesson 1: **Bar charts**

- Record, organise, represent and interpret data in bar charts

Key words
- **statistics**
- **data**
- **interpret**
- **bar chart**
- **horizontal axis**
- **vertical axis**
- **scale**
- **interval**
- **title**

Let's learn

In a **bar chart**, the frequency of the data is shown by the height or length of the bar. The height or length of the bars depends on the scale being used.

To read values between the numbers on the axes, check how the numbers on the axis increase.

Read the marked numbers either side of the bar end or top of the bar to work out the frequency.

Look at the bar chart. With your partner, decide who might have wanted this data and why.

Make up five questions that you can ask.

For example: 'What is the difference in votes between the learners and the school board?' Answer your questions.

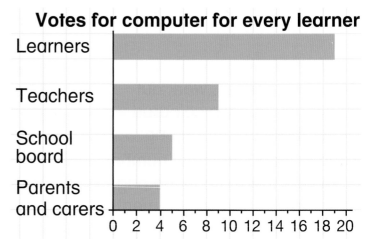

Votes for computer for every learner

Guided practice

Write three pieces of information that this bar chart shows.

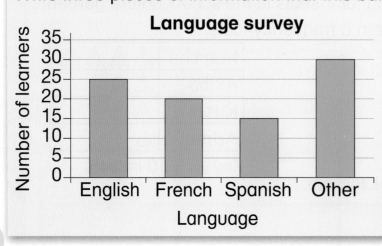

Language survey

25 learners speak English.

5 more learners speak French than Spanish.

10 fewer learners speak Spanish than English.

Statistics and Probability

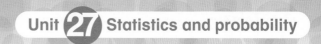

Lesson 2: **Dot plots**

- Record, organise, represent and interpret data in dot plots

Key words
- statistics
- data
- interpret
- dot plot
- title

Let's learn

We can use a **dot plot** to display data.

It is similar to a pictogram, but we use dots instead of symbols.

Tamsin wanted to find out how long her classmates sleep for each night. She asked 30 learners.

These are her results:

8, 7, 9, 8, 10, 5, 6, 6, 7, 8, 7, 6, 5, 9, 8,
9, 6, 7, 8, 7, 5, 6, 7, 8, 8, 6, 5, 7, 7, 8

After collecting the results Tamsin drew a dot plot. She drew a dot for each learner in the column for their number of hours of sleep.

Number of hours slept each night

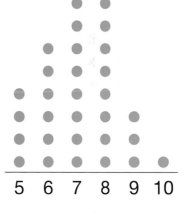

With your partner, make up eight statements from the information in Tamsin's dot plot.

For example, one learner sleeps for 10 hours, and 4 learners sleep for half that number of hours.

Guided practice

Write three pieces of information from the data on this dot plot.

Learners' shoe size

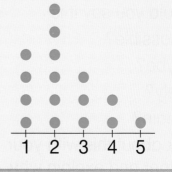

> 6 learners have a shoe size of 2.
> 2 more learners have a shoe size of 1 than 4.
> 16 learners were asked about their shoe size.

Statistics and Probability

Lesson 3: **Chance (1)**

• Use the language of chance

Key words
- chance
- certain
- likely
- maybe
- possible
- impossible

Let's learn

Chance is the possibility that something might happen.

Some things are 'certain' to happen, for example:
November will come after October.

It is 'impossible' for some things to happen, for example:
you will grow an extra leg.

Some things are 'likely' to happen, for example:
we will go to school on Wednesday.

Other things 'may' happen, for example:
if you roll a 1–6 dice, you will roll an even number.

'Impossible' or 'no chance' means something can never happen.
If you roll a 1–6 dice, there is no chance that you will roll a 7.

'Unlikely' means there is a poor chance something will happen.
There is a poor chance that you will win a trip to London.

What does 'even chance' mean?

It means something may happen, but is just as likely not to happen.

If you toss a coin there is an even chance that it will land on heads or tails.

👥 Look at this statement:
I will eat an apple this week.
Would you say it is…
Impossible?
Maybe?
Likely?
Certain?
Talk about this with your partner and explain why you think that.

Guided practice
Write a statement for each of these chance words.

Impossible	I will grow wings and fly.
Maybe	It will rain next month.
Likely	It will be dark at midnight.
Certain	I will go to school sometime this year.

Statistics and Probability

Lesson 4: **Chance (2)**

• Use the language of chance

Let's learn

If you put these cubes in a bag, shake them up and pick one out:

- it is **possible** that you won't pick a red cube
- it is **certain** you won't pick a green cube
- it is **likely** that you will pick a blue cube
- **maybe** you will pick a red or a yellow cube.

Key words
- chance
- certain
- likely
- maybe
- possible
- good chance
- even chance
- poor chance
- no chance
- impossible

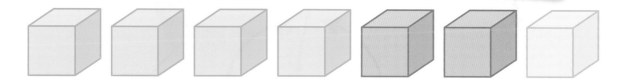

With your partner, think of five things that could happen, but are unlikely to happen.

Now think of five things that might not happen, but are likely to.

Guided practice

Write a statement for each of these probability terms.

No chance — I will *never eat again.*

Poor chance — I will *find $ 100 on the floor.*

Even chance — If I *toss a coin it will land on heads.*

Good chance — I will *eat a hot meal this week.*

Certain — If a bag has *only blue cubes in it, I will pick out a blue cube.*

Statistics and Probability

113

The Thinking and Working Mathematically Star

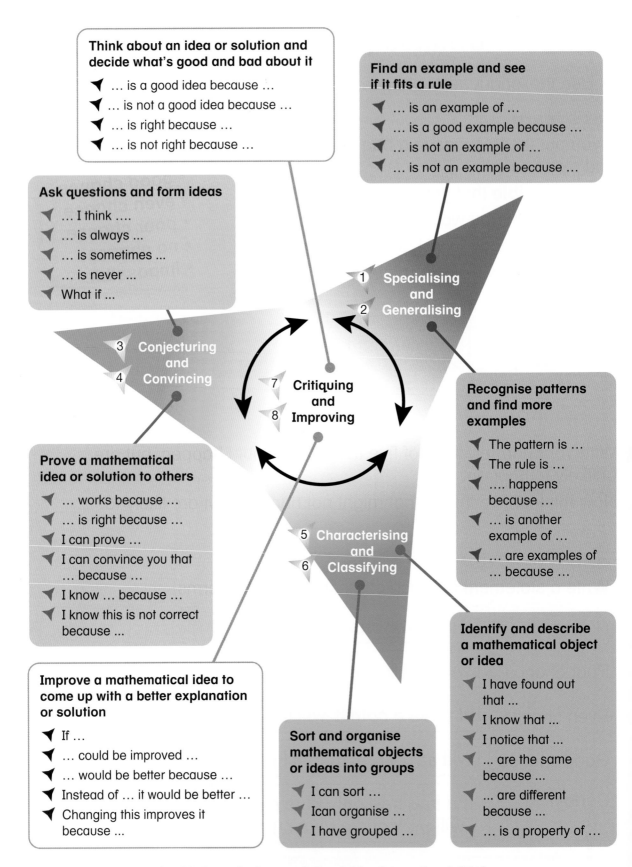

Think about an idea or solution and decide what's good and bad about it
- … is a good idea because …
- … is not a good idea because …
- … is right because …
- … is not right because …

Find an example and see if it fits a rule
- … is an example of …
- … is a good example because …
- … is not an example of …
- … is not an example because …

Ask questions and form ideas
- … I think ….
- … is always …
- … is sometimes …
- … is never …
- What if …

1
2 **Specialising and Generalising**

3
4 **Conjecturing and Convincing**

7
8 **Critiquing and Improving**

Recognise patterns and find more examples
- The pattern is …
- The rule is …
- …. happens because …
- … is another example of …
- … are examples of … because …

Prove a mathematical idea or solution to others
- … works because …
- … is right because …
- I can prove …
- I can convince you that … because …
- I know … because …
- I know this is not correct because …

5
6 **Characterising and Classifying**

Identify and describe a mathematical object or idea
- I have found out that …
- I know that …
- I notice that …
- … are the same because …
- … are different because …
- … is a property of …

Improve a mathematical idea to come up with a better explanation or solution
- If …
- … could be improved …
- … would be better because …
- Instead of … it would be better …
- Changing this improves it because …

Sort and organise mathematical objects or ideas into groups
- I can sort …
- I can organise …
- I have grouped …

The Thinking and Working Mathematically star, © Cambridge International, 2018

Acknowledgements

Photo acknowledgements

Every effort has been made to trace copyright holders. Any omission will be rectified at the first opportunity.
p6 Helen Filatova/Shutterstock; p21 Nor Gal/Shutterstock; p66 Oksana2010/Shutterstock; p71t Natalia
Llaovakaya/Shutterstock; p71b Tarzhanova/Shutterstock; p78t AVN Photo Lab/Shutterstock; p81t Bess Hamitii/
Shutterstock; p81c NataliaPavlovna/Shutterstock; p81bl Thomas Pajot/Shutterstock; p81bc Ymcgraphic/
Shutterstock; p81br Globe Turner/Shutterstock; p82rt Dan Kosmayer/Shutterstock; p82rc Alfmaler/Shutterstock;
p82rb Alhovik/Shutterstock; p88 Mark Brandon/Shutterstock; p98t StudioAz/Shutterstock; p99 WindAwake/
Shutterstock; p112t David Stuart Productions/Shutterstock; p112c AVA Bitter/Shutterstock.